Process Management
in Design and Construction

Process Management in Design and Construction

Rachel Cooper

Adelphi Research Institute for Creative Arts and Science
University of Salford, UK

and

Ghassan Aouad
Angela Lee
Song Wu
Andrew Fleming

School of Construction and Property Management
University of Salford, UK

Michail Kagioglou

Salford Centre for Research and Innovation
University of Salford, UK

Blackwell
Publishing

© 2005 by Blackwell Publishing Ltd

Editorial offices:
Blackwell Publishing Ltd, 9600 Garsington Road, Oxford OX4 2DQ, UK
  Tel: +44 (0)1865 776868
Blackwell Publishing Inc., 350 Main Street, Malden, MA 02148-5020, USA
  Tel: +1 781 388 8250
Blackwell Publishing Asia Pty Ltd, 550 Swanston Street, Carlton, Victoria 3053, Australia
  Tel: +61 (0)3 8359 1011

The right of the Authors to be identified as the Authors of this Work has been asserted in accordance with the Copyright, Designs and Patents Act 1988.

First published 2005 by Blackwell Publishing Ltd

Library of Congress Cataloging-in-Publication Data
Process management in design and construction/Rachel Cooper ... [et al.].
    p. cm.
  ISBN 1-4051-0211-X (alk. paper)
  1. Design, Industrial—Management. 2. New products—Management. I. Cooper, Rachel.

  TS171.P73 2004
  658.5′75—dc22

                                    2004046360

ISBN 1-4051-0211-X

A catalogue record for this title is available from the British Library

Set in 10/13pt Trump Mediaeval
by DP Photosetting, Aylesbury, Bucks
Printed and bound in Great Britain
by MPG Books Ltd, Bodmin, Cornwall

The publisher's policy is to use permanent paper from mills that operate a sustainable forestry policy, and which has been manufactured from pulp processed using acid-free and elementary chlorine-free practices. Furthermore the publisher ensures that the text paper and cover board used have met acceptable environmental accreditation standards.

For further information on Blackwell Publishing, visit our website:
www.thatconstructionsite.com

Contents

Preface

In 1995, Peter Brandon, Professor of Construction at the University of Salford, called me and asked if I was interested in considering construction as a manufacturing process. Quite frankly, I had never heard of it, but the idea was enticing. Having spent many years in research-related design and developments in manufacturing, I was curious to know if knowledge was transferable. This was the beginning of a fascinating journey into the world of construction. Working with colleagues in academia and industry, it soon became obvious that 'process' thinking as we understood it in manufacturing was not common in design and construction. This book is the result of research (funded by the Engineering and Physical Sciences Research Council) investigating and developing new design and development processes for the 'construction product'. The project was led initially by Alfred McAlpine Special Projects and especially by the then Board Director, Dr Richard Baldwin. This was a crucial aspect of the success of the work, since Dr Baldwin had extensive experience in manufacturing and construction, he too could understand the value of improving the design and construction process. Along with Alfred McAlpine, we were also able to engage other partners from across the sector to work with us to understand current issues and to develop a future process. Those partners included further champions, such as Mathew Bacon who was also one of the front guard introducing 'process' management for construction with BAA, and Keith Hamblett doing the same for BT.

As the work developed, many more contributors came from industry – too many to mention, but undoubtedly, without their support, enthusiasm and importantly, their intellectual and practical contributions, the work would not have produced as an appropriate, and therefore targeted, outcome as it did.

Halfway through the development, Professor Tony Thorpe and a team from Loughborough University joined our endeavours and to develop more detail to the original process protocol.

In the eight years since we began our work, 'Process' has been identified by the construction industry as an important issue to address. It is recognised that in order to deliver a 'construction product' on time, on cost and of the highest quality, it is critical to manage the process (and the problems) effectively.

This book provides the context for 'process' thinking. It describes the Process Protocol and the experience of implementing it in practice.

Rachel Cooper

vii

Contributors

Professor Rachel Cooper

Adelphi Research Institute for Creative Arts and Science, University of Salford, UK

Professor Cooper is a Professor of Design Management, and undertakes research in the areas of design and construction process, new product development, design management and socially responsible design. She was Principal Investigator of the EPSRC funded Process Protocol research project and is Principal Investigator of an EPSRC consortium 'Urban Sustainability for the Twenty-Four Hour City', editor of *The Design Journal* and author of four books and over 200 papers based on her research.

Professor Ghassan Aouad

School of Construction and Property Management, University of Salford, UK

Professor Aouad is Head of the 6* rated School of Construction and Property Management, Director of the £3M EPSRC IMRC Centre (SCRI), and leads the prestigious £443 000 EPSRC platform grant (3D to nD modelling). His research interests are in modelling and visualisation, the development of information standards, process mapping and improvement and virtual organisations.

Angela Lee

School of Construction and Property Management, University of Salford, UK

Dr Lee's research interests includes performance measurement, process management, performance-based building and process and product modelling. She has published extensively in both journal and conference papers in these fields. She completed a BA (Hons) in Architecture at the University of Sheffield and her PhD at the University of Salford.

Song Wu

School of Construction and Property Management, University of Salford, UK

Dr Song Wu is a Research Fellow on the 3D to nD modelling project and previously worked on the Process Protocol II project at the University of Salford. He was previously a quantity surveyor in Singapore and China for three years. Song was awarded an MSc in Information Technology in Construction in 2000, and in 2004 completed his PhD at the University of Salford.

Andrew Fleming

School of Construction and Property Management, University of Salford, UK

After gaining an MSc in IT in Property and Construction, Andrew joined the Process Protocol team. He is currently engaged as a Research Fellow on an EPSRC project entitled 'Managing Change and Dependency in Construction'. His research interests are construction process improvement and management, change management, facilities management and development management.

Michail Kagioglou

Salford Centre for Research and Innovation (SCRI), University of Salford, UK

Dr Kagioglou is a Senior Research Fellow and the Manager of the EPSRC-funded Salford Centre for Research and Innovation (SCRI) in the built and human environment. He comes from a manufacturing background, he has published widely and his current research looks at process management and enabling mass customisation capabilities for construction supply chains.

Acknowledgements

The work upon which the Process Protocol project is based took place over a period of six years, between 1995 and 2001, and is continuing in other guises today. The research team was supported with enthusiasm by a group of individuals and companies, without whom the work could not have been done. The group includes John Hinks, Richard Baldwin, Mathew Bacon, George Stevenson, Bob Waterman, Keith Hampson, Paul Jarvis, Nigel Curry, Andrew Waugh, and Trevor Morgan. The companies who partnered us contributed their time and expertise and these include Alfred McAlpine, AMEC, BAA, BIW Technologies, BRE, BT, Capita Property Consultancy, Hammonds Suddards Edge, IAI Client Briefing Domain, IAI Facilities Management and the Waterman Partnership. However, there were many more individuals and organisations that attended the 30 workshops and seminars, and provided insights into a process for the future. The research team were further supported by Daryl Sheath who undertook early work, Stuart Carmichael for his work on Britannia Walk (supported by Tony McCarthy and funded by the DTI) and Jeremy Grammer whose early work on Marconi Lifecycle Management informed ours. From 1999, the research team on the project included colleagues from Loughborough University, including Simon Austin, Andrew Baldwin, Chris Carter, Adam Green, and, most particularly, Tony Thorpe. None of this work would have been achieved without such collaboration or without the funding of the Engineering and Physical Sciences Research Council (EPSRC). Finally, a special thank you to Ginny Spencer for her help in putting the book together and to Andrew Wootton, who designed the first process map upon which all the future iterations were based.

Introduction – Why Process?

The UK construction industry has been under increasing pressure to improve its practices (Hill, 1992; Howell, 1999). It has been continuously criticised for its less than optimal performance by several government and institutional reports such as Phillips (1950), Emmerson (1962), Banwell (1964), Gyles (1992), Latham (1994) and, more recently, Egan (1998). Most of these reports conclude that the fragmented nature of the industry, the lack of co-ordination and communication between parties, the informal and unstructured learning process, adversarial contractual relationships and the lack of customer focus are what inhibit the industry's performance. In addition, construction projects are often seen as unpredictable in terms of delivery time, cost, profitability and quality, and investment into research and development is usually seen as expensive when compared to other industries (Egan, 1998; Fairclough, 2002).

According to Howell (1999), the 'inefficiency' of the industry has tended to be the way of life. This may be due to the fact that none of the reports, apart from Latham (1994) and Egan (1998), have been sufficiently acted upon. As Latham (1994) points out, '... some of the recommendations of those reports were implemented ... but other problems persisted, and to this day, even the structure of the industry and nature of many of its clients has not changed dramatically'. Therefore, Latham (1994) suggests using manufacturing as a reference point and Egan (1998), in his *Rethinking Construction* report, recommends process modelling as a method of improvement.

The transfer of practices and theories from other sectors, as suggested by Latham (1994), has been a constant subject of discussion since the publication of his report. Some construction practitioners are adamant that their industry is unique and that the transference of principles cannot be adopted wholeheartedly. Ball (1988) highlights some of the arguments most commonly used to distinguish construction from other industries:

- The one-of-a-kind product.
- The spatial fixity of buildings.
- On-site production.
- The effect of land price on design and construction possibilities.
- The requirement for long life expectancy.
- The inexperience of clients.
- The merchant/producer role of companies.
- The overwhelmingly domestic industry.
- The masculine stereotype of the workforce.

1

- The long cycle from design to production.
- The high cost of the projects.
- The amplified reaction to economic crises.
- The labour-intensive production.
- The fragmented nature of the industry.

In contrast, there are also many practitioners and academics who believe that the construction industry has much to learn from manufacturing. Howell (1999) goes so far as to suggest that this learning could be a two-way process: manufacturing could learn from construction in areas such as project-based management: and construction could learn, from manufacturing's developed and developing solutions, to improve its competitiveness.

According to Koskela (1992), Love & Gunasekaran (1996) and Kornelius & Wamelink (1998), manufacturing has been a constant reference point and a source of innovation in construction for many decades. Solutions that have been recommended to help overcome the problems of construction include industrialisation (i.e. prefabrication and modularisation), computer-integrated construction, robotics and automated construction (Koskela, 1992; Love & Gunasekaran, 1996; Kagioglou *et al.*, 1998a). However, their implementation in manufacturing is far advanced in comparison to the construction industry. Koskela (1992) believes that the underlying theories and principles of manufacturing should be harnessed to deliver the full benefits to construction rather than the 'technological solutions'.

The realisation that the construction industry might not be as unique as was traditionally thought has initiated new research in recent years. In particular, this has led to the development of the 'Construction as a Manufacturing Process' research fund under the Innovative Manufacturing Initiative (IMI) sector of the Engineering and Physical Sciences Research Council (EPSRC, 1998) to continue and expound upon current thinking. (This book is based on research funded under that initiative.)

It now appears that a new phenomenon is being steadily exploited within construction companies alongside the new technologies taken from manufacturing. It is based upon the development and use of fundamental core processes to improve the efficiency of the industry, with great emphasis upon the basic theories and principles underlying the design and construction process. Egan (1998) highlighted this factor by reporting that due to the fragmented nature of the construction industry very little work had gone into process modelling. Manufacturers are accustomed to taking a process view of their operations; they usually model both discrete product activities and holistic high-level processes for both internal and external activities. In particular, there has been a growing volume of research focusing upon the consolidation of the just-in-time (JIT) and the total quality management (TQM) philosophies, with an array of other practices such as total productive maintenance, visual management and re-engineering (dos Santos *et al.*, 1999). Investigations by construction practitioners and academics alike have now sought to develop the content and structure of the core ideas underlying these theories, namely world-class manufacturing, agile production and lean production (Schonberger, 1996; Gilgeous & Gilgeous, 1999). This has led to a range of corresponding practices, for instance, world-class construction, agile construction and lean construction, as it is believed that process improvement in the construction industry may well be a significant strategy for getting the right product to the right market at the right time, cost and quality (Pheng & Tan, 1998).

As the construction product has in most instances been a 'one-off', much emphasis has been placed on project management. Yet in effect the industry is concerned with the design and development of a building product and should look to manufacturing for reference on how to manage the design and development process. This book will examine the manufacturing perspective and will illustrate how it can be applied to design and construction through the use of a case study in the development of a Generic Design and Construction Process Protocol. It will also consider the use of the techniques and technologies available to support the process and the issues relating to their implementation on projects.

1

The Product Development Process

'Product development is fundamental to stimulating and supporting economic growth for organisations and for wealth generation in many industrialised nations . . . product development is a
strategic process, and product development and design activities
are powerful corporate tools.'

Bruce & Biemans, 1995.

In order to overcome the barriers within construction as identified in the
Introduction, it was suggested that construction should be viewed as a
product development process. It is therefore important to understand current thinking on new product development (NPD). This chapter uses the
product development process in manufacturing as a reference point for
defining and understanding the design and construction process (see Fig.
1.1). The importance of new product development is discussed together
with the activities and models used to illustrate it. Having considered
briefly the history of construction, its project-based orientation and the
existing models of the design and construction project process, the chapter
will conclude with an explanation as to why a holistic product development
view of construction is necessary.

Product development in manufacturing

If the world were stable there would be no need to change business operations and methods or to understand what has changed and what works
well. However, firms operate in dynamic environments, not stable ones,
and both external competition and internal environments evolve over time.

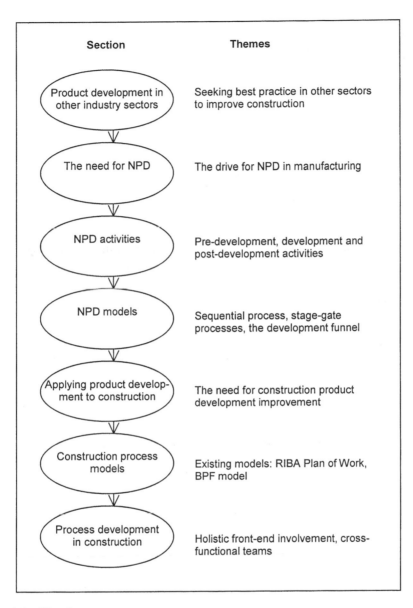

Section   **Themes**

Product development in other industry sectors — Seeking best practice in other sectors to improve construction

The need for NPD — The drive for NPD in manufacturing

NPD activities — Pre-development, development and post-development activities

NPD models — Sequential process, stage-gate processes, the development funnel

Applying product development to construction — The need for construction product development improvement

Construction process models — Existing models: RIBA Plan of Work, BPF model

Process development in construction — Holistic front-end involvement, cross-functional teams

Fig. 1.1 Chapter map.

In response, processes must also continuously adapt to enable those firms to remain effective and profitable through the changing conditions (Moran & Brightman, 1998). Therefore organisations wishing to undertake improvements in productivity, quality and operations need to reconsider their working practices (Elzinga *et al.*, 1995). Katzenbach (1996) reports that organisational change is becoming everyone's problem and that customers require it, shareholder performance demands it and continued growth depends upon it. Customer/client awareness and expectations have increased in terms of quality and value for money. The manufacturing industry has been developing new approaches to NPD since the 1970s by modelling and improving its processes. Egan (1998) supports this drive and suggests that the construction industry could also improve its performance by modelling its processes.

The need for new product development (NPD)

In a dynamic economy, developing and introducing new products is essential for a company's survival (Schmidt, 1995) and the successful management of new products has become both a necessity and a way of life (Sarin & Kapur, 1990). A number of studies have indicated that companies rely on new products to generate profits (Baker, 1983) and will continue to do so, to a greater extent, in the future (Thomas, 1993). Ames & Hlavacek (1984) indicate an increase in profits contribution from new products: from nearly 23% in the period 1978–81, it increased to 32% in the period 1981–86. Booz, Allen & Hamilton (consultancy practice) surveyed 700 companies in 1982 and reported that 31% of the companies' profits would come from new products over the next five years. Moreover, in the USA in the year 2000, 50% of company profits came from new products that were five years old or less (O'Connor, 1986).

New product development (NPD) is a necessary risk that companies must undertake. Technological developments, shorter product life cycles, the complexity of products, increasingly changeable market demands, customers who demand 'the best', and stronger and more global competition mean that companies face a limited space in which to succeed (Ross, 1994; Trygg, 1993; Oh & Park, 1993; Inwood & Hammond, 1993; Gupta & Wilemon, 1990).

NPD is a critical means by which the whole organisation – the business as well as the employers – can adapt, diversify and, in some cases, reinvent the firm to match evolving market and technical conditions (Schoonhoven *et al.*, 1990). Brown & Eisenhardt (1995) suggest that, although technical and market changes can never be fully controlled, proactive product development can influence the competitive success and renewal of organisations.

Since the 1970s, and particularly since the mid-1980s, the literature on new product development has grown very large. Many studies have been undertaken to determine critical success factors in NPD (Cooper, 1992; Clark & Fujimoto, 1991; Zirger & Maidique, 1990; Cooper & Kleinschmidt, 1987b; Rothwell *et al.*, 1974; Rothwell, 1972). Rothwell (1972) identified a number of success factors related to the individual activities involved in the NPD process, and concluded that the way in which those activities are performed needs to change in order to increase competitiveness, success and survival rate. The sequence and relevance of those activities, among themselves and in relation to the rest of the organisation's activities, form the NPD model. Cooper (1994) defines the model, or new product process, as '...a formal blueprint, roadmap, template or thought process for driving a new product project from the idea stage through to market launch and beyond'.

NPD activities

It is widely accepted that in order to move a new product idea through to production and on to final launch in the marketplace a number of activities must be performed (Utterback, 1971). Initiated by the identification of a need or the adoption of an idea, a number of preliminary evaluations are carried out. Further detailed technical development follows and finally, after a series of company and market tests, the finished product is launched onto the market (Crawford, 1994). The way in which these activities are performed has been, and still is, a subject of research and has resulted in a number of new product development process models.

The number of stages involved in an NPD process ranges from six (Booz, Allen & Hamilton, 1982) to as many as 13 (Cooper & Kleinschmidt, 1986) and many firms frequently omit, either intentionally or accidentally, some of these activities when developing new products (Dwyer & Mellor, 1991; Sanchez & Elola, 1991). Generically, the NPD activities can be separated into three broad main categories: the pre-development activities, the development activities and, finally, the post-development activities (Cooper & Kleinschmidt, 1988).

Pre-development activities

The sources and ultimate users of the information needed for the pre-development activities are provided from within the company (research and development, marketing, manufacturing, sales and management) or from outside (customer needs and requirements) (Rochford & Rudelius, 1992). In a logical order, the first activity of the NPD process is idea generation or establishing the need, followed by a number of preliminary market, technical, financial and production assessments (Marquis, 1972). Baker *et al.* (1983) defines the idea or need as '...a potential proposal for undertaking new technical work which will require commitment of significant organisational resources', and idea generation as the '...coming together of an organisational need, problem or opportunity with a means of satisfying the need, solving the problem, or capitalising on the opportunity'.

A number of preliminary investigations regarding technical compatibility and capability of the company, financial attractiveness in terms of return on investment (ROI) or pay-back period (PBP) and market assessments are the catalysts in making a decision regarding future development of the product. Juran (1988, 1989) refers to this stage as the translation of the customer's (internal or external) needs into the company (internal or external). However, Cooper (1988) found that many firms face problems in this translation stage, the result being a final product that is '. . . not quite what the customer wants, or [that] lacks differentiation from existing products'.

The deliverables (outputs) of the pre-development activities can be presented in terms of product design specifications (PDS), resource allocation and process needs, market share and post-development marketing and company policy. The design brief or PDS can be formulated from a variety of sources such as stakeholder requirements, market research, legislation, benchmarking of the 'best' manufacturer (competitor) of analogous products, reports, proceedings and the identification of market gaps together with statistical data and other tools (Pugh, 1991). The PDS provides the functional characteristics of the product including aesthetics, size and weight, operating conditions and environmental factors. There are often many solutions that satisfy a specific functional requirement. A premature decision at this stage of the development could greatly affect the overall cost and complexity of the product (Smith & Reinertsen, 1991) and, since every stage of the NPD is affected by the PDS, it is crucial to the overall success of the product development (Wnuk, 1990).

Crawford (1984) suggests that an important tool at this stage of the NPD is the development of a product protocol. It is very similar to the PDS, as described above, and in addition to establishing the business case, it defines the target market, the benefits the product will/could deliver and how the product will be positioned in the market.

Development activities

The development activities relate to the actual physical development of the product and the process. The task of the designer or the design team (multi-functional and multi-disciplinary) is to create the design both of the product and of the manufacturing process. This can yield the desired production quantity at a price that allows the company to sell the product for a profit (Nevins & Whitney, 1989). Under the heading of 'development activities' testing and validation of the product can also be included, which differs from the construction industry's approach.

A single or, in many cases, a number of conceptual designs are fully developed (Ward *et al.*, 1994). Also one or more prototypes are developed (something that the construction industry cannot enjoy due to the nature of its product) to create a physical model of the product so that its performance and functionality can be examined (Pannesi, 1994). The prototype is tested in-house (alpha test) and by the customers (beta tests) (Crawford, 1994). The manufacturing process is tested in a limited trial, or pilot production, to prove the capability of the production process and to determine more precise production costs and throughputs. Market tests and financial analyses will determine if the project is still viable in terms of expected market share revenues and cost data (Cooper, 1993).

Post-development activities

The post-development activities are related to the final launch of the product into the market place, the marketing approach, the after-sales support and finally a review of the project's effective implementation, any mistakes made and the performance of the activities involved. The last of these will form the basis for future product and process improvements through customer (internal and external) feedback in terms of future requirements or current faults; and NPD process improvements in terms of identifying deficiencies in the process model or in the implementation of the individual activities (Holmes, 1994).

The lessons learnt can also be used for the development of new products. This is an area that has not been considered carefully in the past, resulting in the same mistakes being repeated in future projects.

NPD models

Depending on the number and nature of the activities of an NPD process adopted by individual firms, NPD models can be represented in various ways. They may incorporate the product policy of the company prior to development and/or the lessons that can be learned from a successful or unsuccessful development. Often, though, an NPD model only represents the particular projects, and company policy prior to and after the completion of the project is not included. From a historical point of view NPD models can be classified into three main streams: sequential, overlapping and stage-gate phase review.

Sequential approach

In a sequential or serial approach to NPD the development moves through different, almost mutually exclusive, phases as shown in Fig. 1.2, in a

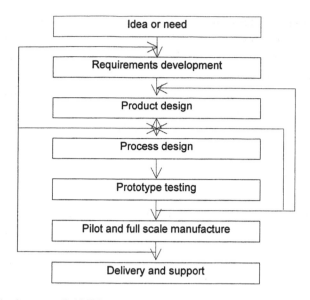

Fig. 1.2   Typical sequential NPD process.

logical step-by-step fashion (Imai *et al.*, 1985). This approach has been exemplified by the National Aeronautics and Space Administration's (NASA) phased program planning (PPP) model. Concept decisions, product design and testing are performed prior to manufacturing system design, process planning and production (Stoll, 1986).

The development proceeds to the next phase/stage only after all the requirements of the preceding phase are satisfied. In each succeeding phase of the project, new and different intermediate results are created with the outputs of one phase forming major inputs to the next (Coughlan, 1991). The design activities are isolated (the 'over the wall' effect) from the realities of the issues facing test, manufacturing, quality and service as shown in Fig. 1.3. This is very much like the traditional construction processes which will be described later in this chapter.

This type of approach results in a product development process that is essentially linear with some very hard breaks between the phases of the

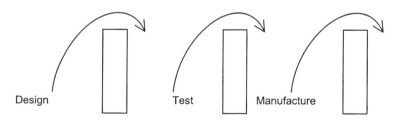

Fig. 1.3   Sequential 'over the brick wall' approach.

process (Harkins & Dubreuil, 1993). It is often a result of the organisational structure of the company: highly structured and functionally organised into departments of individual principles such as engineering, R&D, marketing and sales. Each function is expected to play a specific and limited role in any phase. For example, the engineering department designs the product, manufacturing makes it and marketing sells it (Hayes *et al.*, 1988). If a feature or an important element of the product is omitted at the early stages, this fault will probably ramify as it moves through the different stages resulting in an obsolete product (Oakland, 1995). The designs are passed from design to manufacturing and *vice versa*, resulting in long lead times, late product launch, increased development costs, lack of information flow and lack of flexibility for change in the process (Desa & Schmitz, 1991; Turino, 1990; Stauffer, 1988; Putnam, 1985).

Apart from the limitations and disadvantages of the sequential approach, it offers high staff utilisation in departments and is favourable for breakthrough projects that require revolutionary innovation. It is also suited to very big projects where the sheer number of personnel involved limits extensive communications between the members of the team, and where product development is masterminded by a genius who creates the invention and hands down a well-defined set of product specifications (Takeuchi & Nonaka, 1986).

However, a decision concerning product design tends to have a number of significant manufacturing and non-manufacturing impacts upon the life cycle of the product (Dowlatshahi, 1994). A study at Rolls-Royce revealed that design determined 80% of the final production cost of 2000 components (Gorbett, 1986). Huthwaite's (1988) study revealed that product design activities are responsible for only 5% of a product's cost; they can, however, determine 75% of all manufacturing costs and 80% of a product's quality performance. The Ford Automotive Company has estimated that among the four manufacturing elements of design, material, labour and overhead, 70% of all production savings stem from improvements in design (Cohodas, 1988) and 70% of the life-cycle costs of a product, in terms of materials, manufacture, use repair and disposal of the product, are determined at the design stage (Nevins & Whitney, 1989). Turino (1990) found that in the electronics industry the costs of making changes in the design of a product are far greater at the later stages of the new product development process, as shown in Table 1.1.

It became apparent, therefore, that the next step of NPD will need to incorporate areas such as manufacturing, assembly, quality, R&D and marketing at the design stage.

Table 1.1 Typical costs of design changes
in the electronics industry.

Design changes made	Cost
During design	$1,000
During design testing	$10,000
During design planning	£100,000
During pilot production	$1,000,000
During final production	$10,000,000

Stage-gate processes

NASA's PPP (phased project planning) process, which is often referred to as phase-review process (Rosenau & Milton, 1990), had a number of disadvantages as discussed earlier. However, Milton & Rosenau (1988) suggested that when the phase-review process is executed by CFTs (cross-functional teams) it offers a number of benefits such as reducing risk, easing the task of setting goals toward completing each phase, and improving focus on a particular phase.

One such process gaining wide acceptance (O'Connor, 1994) is known generically as 'stage-gate', and is illustrated in Fig. 1.4 (Cooper, 1990). It is presented as a series of gates and stages the number of which can vary from typically four up to seven (Cooper, 1993) depending on the organisation using it. Each stage represents a number of activities that need to be performed and the information that needs to be gathered to progress the project to the next gate. Unlike PPP, the stages do not represent single functional activities in the organisation but are multi-functional, involving a number of people from different departments relevant to the activities. Gates represent decision or 'go/kill' points which specify a set of criteria or 'deliverables' that the project must or should meet in order to proceed to the next stage of development. The gates serve as quality control checkpoints and are usually controlled by senior managers from different functions who own the resources required by the project leader or team.

Stage-gate processes have been found to reduce development time, produce marketable products and optimise internal resources by eliminating projects that are not promising (LaPlante & Alter, 1994; Anderson, 1993; Cooper & Kleinschmidt, 1991). There are, however, a number of disadvantages and weaknesses associated with stage-gate systems. Cooper & Kleinschmidt (1991) found that in some companies management indicated that the process takes too long to learn and perform.

O'Connor (1994) found that process customisation and a process management superstructure are necessary for a large firm to implement and

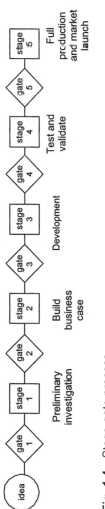

Preliminary
investigation

Build
business
case

Development

Test and
validate

Full
production
and market
launch

Fig. 1.4 Stage-gate process.

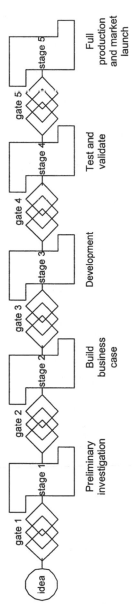

Preliminary
investigation

Build
business
case

Development

Test and
validate

Full
production
and market
launch

Fig. 1.5 Third-generation NPD process.

manage a stage-gate product development process. There are generally six deficiencies of the stage-gate system (Cooper, 1994):

- Projects must wait at each gate until all tasks have been completed. Thus, projects can be slowed down for the sake of one activity that remains to be completed.
- The overlapping of activities is not possible.
- Projects must go through all stages and gates when in some circumstances it might be quicker, especially for small firms, to eliminate or bypass some activities.
- The system does not lead to project prioritisation as it was originally designed for single projects (Griffin & Hauser, 1996).
- Some new product processes are very precisely defined, accounting for minute details in the process and making the process hard to understand, manage and learn.
- Sometimes the system tends to be bureaucratic, making the process too slow.

To overcome these deficiencies of the stage-gate system Cooper (1994) suggested the 'third generation new product process' which illustrates how the NPD model might look in the future (Fig. 1.5). The main characteristic of the new process is the overlapping of the stages. Go/kill decisions are delayed to allow for flexibility and speed and the previously 'hard' gates are presented as 'fuzzy' gates, which can be either conditional or situational.

Conditional gates relate to a 'go' decision made subject to a task being completed at a specified point in time and the results of that task indicating that it is still a good project. Situational gates refer to a 'go' decision being made when the information from a task that is not yet complete is not vital enough to halt the project. Because of the overlapping of stages and for the sake of flexibility, decision-making authority will shift away from senior management and more towards the team and the team leader. The process does, however, remain sequential between consecutive stages as stages cannot be bypassed or eliminated from the process.

Successful NPD has always been the primary objective of any firm developing new products. The criteria that define success differ between firms (Hart, 1996) and they form the basis on which firms benchmark their NPD activities. Throughout the years many models and roadmaps have been created to demonstrate, organise and manage the NPD activities in whatever ways companies have seen fit. Figure 1.6 illustrates the drivers and respective approaches that have been pursued by new product development organisations. Essentially the drivers relate to flexibility and speed, focus and control, and provide a consistent roadmap for managing the new product development process. There is no simple or single solution

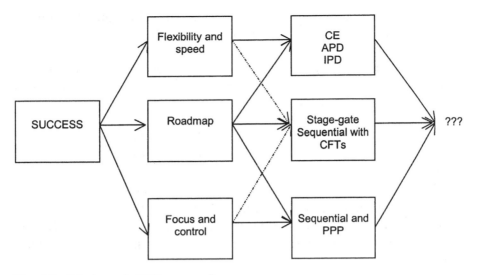

Fig. 1.6   What now in NPD process?

to new product development but a set of possible tools and philosophies to be combined together (Smith & Reinertsen, 1991). What works for one company might not work for another for a variety of reasons such as product and process complexity, nature of products (high-tech or low-tech), duration of projects, markets targeted and the nature of the company itself (technology or market orientated).

The approaches described so far have a number of advantages and disadvantages associated with them and a good balance between them should offer advantages that none of them can offer alone. There needs to be a balance between the approaches that offer flexibility and speed and those that offer focus and control. In general there are three main activities that should be carried out by companies developing new products:

(1)   Select the 'right' new product idea.
(2)   Select the 'most effective' process for the product to enter the market 'successfully'.
(3)   Learn from the 'good and bad points' of the product and process selection.

The development funnel

Wheelwright & Clark (1992) suggest a solution that incorporates all of the above issues in the development funnel, as shown in Fig. 1.7. The first phase of the development funnel represents the concept development and idea generation for potential product/process efforts. Screen 1 represents an

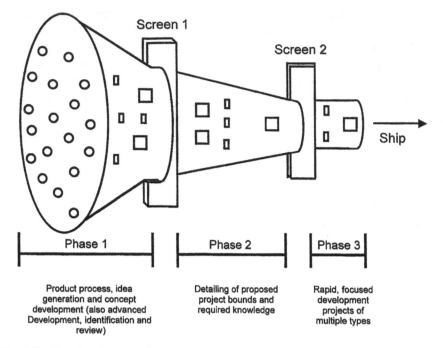

Fig. 1.7   The development funnel.

evaluation point performed by a mid-level group of managers (peers) identifying the information needed for a go/kill decision at screen 2. As part of
screen 1 ideas should be checked for their fit with technology and product
market strategies and their appropriateness as an application of the firm's
development resources. Screen 2 operates as a gate, as seen for the stage-
gate process, operated by senior management. A review of the product and
process is used to select those projects that will be fully developed.

The approach to the development process should be customer and market
focused, and should include cross-functional teams. The procedures,
phases, and rules used should be streamlined, appropriate and adaptable.
The results of the project should be used as learning agents for future
development. The development funnel itself, however, can pose significant
constraints on resources as not all companies have the capability to allocate
resources for the conceptual development of a large number of projects.

Applying product development to construction: historical background

In order to understand product development in construction it is important
to set it in the context of the history of construction. This historical

background includes unique social and political relationships that through their development have contributed to the apparent fragmentation of today's design and construction process.

Early forms of construction procurement are not well described. However, what is clear is that the formality of the construction project process has varied according to the accepted social hierarchy of monarchy, church and public. According to Colvin (1975), works undertaken on behalf of the monarchy or church would have involved the management and administration of construction projects by a local constable, sheriff or clergyman who would have overseen and co-ordinated the general building trades. In contrast, public construction projects were carried out much more informally. In this sense, the introduction of procurement as a distinctive administrative and management structure for construction projects may be associated with the 1500s. At this time Henry VIII's 'passion for building', as described by Colvin (1975), vastly increased the construction workload and this, it would appear, necessitated the development of structured administration both on and off site.

This period is worth noting as through the 1600s and 1700s a noticeable change in social structure led to the creation of a landed aristocracy. Demand for construction shifted significantly from the monarchy to the new possessing class who felt a need to emulate their rulers in building but had no means of doing so. Thus the structure of procurement altered to meet this new demand. Effectively, specialist client's agents now filled a gap in the structure of construction management, acting as intermediaries between the aristocratic client and the building trades. Those who had previously serviced the monarchy provided this intermediate service. The roles provided by the skilled artisan and master craftsman were now demanded by an 'independent' market. It is here that the first signs of a construction management service are identifiable.

In the 1700s the Bank of England was established and with it, right through into the 1800s, the UK economy gradually began to grow. Trade, both domestic and overseas, produced a quantitative increase in demand for building stock of all kinds. Agrarian and industrial revolutions placed further demands on the construction industry, as did the construction of national transportation systems such as canals and railways. Many master masons and artisans found themselves able to specialise further. The engineering and architectural functions became distinct from one another, as did the various surveying functions.

Qualitative changes also occurred during this period as the spirit of commercialism filtered through society and many of the population benefited from a greater distribution of wealth. More and more, clients wished to demonstrate their social position through the buildings in which they lived and worked. Buildings were noted for being prestigious and imposing and produced a demand for architecture that undoubtedly assured the architectural profession's position in the construction system. In construction, the management functions became increasingly defined and many, especially architects, began to specialise and form professional institutions.

By the mid 1800s the UK was experiencing the effects of the mass quantitative surge in population which had occurred during the industrial revolution. This, plus the effects of rural–urban migration, created a demand for housing on a quite unprecedented scale. Such demand led to much standardisation in both form (for example, terracing) and production in house building. General contracting (or contracting in gross) enabled many to profit from the increased workload and in many speculative developments design and build forms of procurement were adopted to facilitate profitable production.

Powell (1980) suggests that building generally in the late 1800s was a 'manifestation of Victorian commercial spirit'. This, in addition to Bowley's (1966) assertion that in the early 1900s the architect–surveyor/engineer–builder hierarchy was the 'only really respectable' form of organisation, would therefore seem to apply in what was becoming an increasingly demarcated society. By the early 1900s the commonly termed 'traditional' hierarchy of construction organisation was established. Industrial and technological developments were rapid and, through colonialism and other overseas links, trade boomed.

World War I left many regions of the UK in dire need of rebuilding work for all types of building stock. The welfare state programme of public expenditure began at this time and continued after World War II, altering the level of real wealth in the community. Planning controls were introduced, social services were extended and, in real terms, a programme of redevelopment commenced. The early years of the twentieth century had seen a qualitative change in the construction market. Clients were increasingly commercially orientated although there was still great demand for industrial buildings. Resources such as land and building materials were scarce or their use was controlled. To many it felt that these new demands required a change in the conventional approach to procurement. Negotiated forms of procurement were introduced in the inter-war years and, as time progressed, management forms of procurement were experimented with and adopted.

Post 1944 industry has experienced continued calls for integration among the management functions and recognition of the value that each member of the construction management organisation can contribute. The corresponding increase and variation in procurement systems reflects this.

A process of experimentation and adaptation continued through the mid-1900s and by the 1960s many distinct forms of alternative procurement were in use. Their adoption was necessitated by the change in the type of client demand that had occurred post World War II. All forms of industry were increasingly concerned with the concept of management. The world economic 'crisis' of the 1970s intensified this and management was seen more and more as the key to efficiency and effectiveness in all types of business. Client requirements were becoming more detailed and produced a change in demand that required a different approach from the suppliers of construction. Despite the trend towards management, the industry remained fragmented with a large 'trail' of small sub-contractors and suppliers and thus fragmented approaches to managing the design and construction process (DCP).

In 1959 the United Nations defined the building (project) process as '... the design, organisation and execution' of a building project that has come to be recognised as '... normal practice in any country or region ... it is characterised by the fact that all operations follow a set pattern known to all participants in the building operation' (United Nations, 1959). However, this description is not so succinctly true today. The nature of the design and construction process has grown in complexity since the 1950s, leading to an increased number of actors in the project. According to Howell (1999) variation occurs in every project process. Indeed there now arguably appears to be no 'standard project process' of construction, either within a country or within a region, and no clear indication of the roles and responsibilities of the project participants.

The term 'the traditional building process' today usually refers to the practice where, upon perceiving a need for a new facility, a building client approaches an architect/engineer to initiate a process to design, procure and construct a building to meet his/her specific needs. The process, in turn, almost invariably consists of the project being designed and built by two separate groups of disciplines who collectively form a temporary multi-organisation for the duration of the project: the design group and the construction group (Mohsini & Davidson, 1992). Typically, the design group is co-ordinated by an architect/engineer. Depending upon the circumstances of the project at hand it may also include other design professionals and specialists such as engineers, estimators and quantity surveyors. The principal function of this group is to prepare the design specifications of the work and other technical and contractual documents. The construction group, on the other hand, is usually co-ordinated by the main contractor and consists of a host of sub-contractors and suppliers/manufacturers of building materials, components, hardware and subsystems. This group is primarily responsible for the construction of the building project.

The two groups typically do not work coherently together (Kagioglou *et al.*, 1998a). The design activities in construction are usually isolated from the realities of the issues facing production as each function is expected to play a specific and limited role in any phase, thus contributing to the industry's problems as highlighted by the many governmental and industrial reports (Phillips, 1950; Emmerson, 1962; Banwell, 1964; Gyles, 1992; Latham, 1994; Egan, 1998). This factor has contributed to the problems in construction of poor supply chain co-ordination and fragmented project teams with adversarial relationships (Mohsini & Davidson, 1992). Following these reports new procurement routes have emerged to ease the problems and some of these are outlined below:

- *Construction management:* a construction manager is employed alongside the design team as a project is initiated. The manager defines and manages the work packages to allow construction work to commence long before the design of the entire project is complete. However, costing of the project cannot be accurately forecast until all the packages have been let (Kagioglou, 1999).
- *Management contract:* design by the project sponsor's consultants and construction overlap. A management contractor is appointed early to let elements of work progressively by trade or package contracts. All contracts are between the management contractor and the trade contractors. As with construction management, the final cost can only be accurately forecast when the last package has been let (Carty, 1995).
- *Design and manage:* similar to the management contract but here the contractor is also responsible for detailed design or managing the detailed design process (Fryer, 1997).
- *Design and build:* contract for both design and construction of a project by a single contractor for a lump sum price (Carty, 1995).

Construction project process models

The emergence of various forms of procurement routes is perhaps the most significant attempt that the UK construction industry has made to improve its services (Masterman, 1992). According to Hibberd & Djebarni (1996) the concept of procurement raises awareness of the issues involved in challenging generally accepted practices and establishing strategies, thus the need to consider new approaches to the design and construction process. Latham (1994) argues that reducing variations in the project process will improve performance and make significant cost savings. The fundamental benefit of such an improved design and construction project process should be to optimise predictability (Kagioglou, 1999). This can only be ensured when a truly co-operative project environment exists. The project process should look to facilitate team working and effective communication between participants (Kagioglou *et al.*, 1998a). Further, information technology (IT) can assist the attainment and maintenance of a new project process if its operation and the relationship between the parties is sufficiently prescribed and detailed.

Unfortunately there does not exist a means by which to reduce variation in construction in order to improve performance (Latham, 1994). The current perception is that flexibility is difficult within the process because the supply chain changes for every project and relationships are dynamic. Despite the lack of a 'standard' project process there are several well-recognised models of the construction process, namely the Royal Institute of British Architects (RIBA) *Plan of Work* (first published in 1964) and the British Property Federation (BPF) manual (1983). (Other process models that have been recently developed tend to only replicate a particular aspect of the design and construction process).

RIBA Plan of Work

The *RIBA Plan of Work* (RIBA, 1997) was originally published in 1964 as a standard method of operation for the construction of buildings and it has become widely accepted as the operational model throughout the building industry (Kagioglou *et al.*, 1998a). However, it was designed from an architectural perspective. This has in some ways restricted its applications to Joint Contracts Tribunal components and it is not generic enough for wide construction works.

The plan (see Fig. 1.8) represents a logical sequence of events that should ensure that sound and timely decisions are made during the course of the project. It suggests that all the decisions, set out or implied, have to be taken

Pre-design	A B				
Design		C D E			
Preparing to build			F G H		
Construction				J K L	
Post-construction					M

Stage A: Inception
Stage C: Outline proposals
Stage E: Detail design
Stage G: Bills of Quantities
Stage J: Project planning
Stage L: Completion

Stage B: Feasibility
Stage D: Scheme design
Stage F: Production info
Stage H: Tender action
Stage K: Operations on site
Stage M: Feedback

Fig. 1.8   *RIBA Plan of Work.*

or reviewed (RIBA, 1997) and it is anticipated that the model will only need slight adjustments depending upon the size and complexity of the project. The project progresses from inception to feedback, i.e. from stages A to M, in a linear fashion requiring the completion of one stage before proceeding to the next. However, the design and construction process is essentially not linear and cannot be viewed in such a functional fashion. Moreover, this sequential flow only aids the hard breaks between the organisational structure of the industry and contributes to the problems of fragmentation and poor co-ordination and communication between project team members (Sheath *et al.*, 1996), as highlighted earlier by many governmental and institutional reports (Phillips, 1950; Emmerson, 1962; Banwell, 1964; Gyles, 1992; Latham, 1994; Egan, 1998).

British Property Federation (BPF) model

The formation of the British Property Federation (BPF) model was a direct result of the growing concern at the increasing problems within the construction industry, notably poor design, inadequate choice of materials and poor supervision of the works combined with a lack of representation of the private sector client (Kagioglou *et al.*, 1998a). The model was intended for use by all those involved in a construction project, i.e. client, design consultants, contractors, subcontractors and suppliers, which was where the *RIBA Plan of Work* was lacking. It highlights the formal and informal relationships between these parties and aims to provide the client with value for money from the construction process by dividing the design and construction process into five stages (British Property Federation, 1983):

- Concept.
- Preparation of the brief.
- Design development.
- Tender documentation and tendering.
- Construction.

The model sets out to be flexible and allows the client to make a decision as to whether to continue with the project at the end of each stage. Furthermore, the model can determine the actual position of the dividing line between stages, outlining when to make that decision. Although the model has not been widely implemented, which may be due to its close link with repetitive house building projects, it has many advantages over the 'normal' methods of design and construction such as (British Property Federation, 1983):

- It produces better buildings more quickly and at lower cost.
- It removes the overlap of effort between design team members.
- Through more thought at the initial stages of the project fewer variations are needed when on site, resulting in fewer delays, a lower cost and improved performance by the design team.

Process development in construction

The success of construction projects largely hinges on both the efficacy of the project process and the team enacting the work (Sidwell, 1990). The construction team is a living organism usually formed from various organisations for the temporary duration of the project (Lingle & Schiemann, 1996). Although they have different priorities and capabilities they are expected to work cohesively together from the outset. However, cultural issues between project team members have often been cited in the literature as limiting the project success (Sidwell, 1990); team-building exercises are commonly introduced to counteract this. In addition, the number and subsequent variations of design and construction project processes exert a much deeper influence on the efficacy (Howell, 1999). It was stated earlier in this chapter that there existed no standard project process (Latham, 1994) that is succinctly followed. This gives a clear indication of the problem facing construction: how are professionals meant to instinctively organise themselves into a team-working environment when the process is full of infinite variation in that their roles and responsibilities vary from project to project? Latham (1994) and Egan (1998) suggested that learning from manufacturing and process modelling would aid project success. In particular, two key lessons must be considered: holistic front-end involvement (taken from the link between NPD pre-development, development and post-development activities) and the subsequent use of cross-functional teams. The next chapter will discuss tools and techniques that can be used to improve the design and construction process.

Holistic front-end involvement

The development of the construction process has left a sharp antithesis between the architect and the contractor (Harvey, 1971). The barriers have generated attempts to change the situation though these have tended to concentrate upon altering the structures and processes to improve information flow and to reallocate risk through new mechanisms such as design and build, prime contracting, partnering, management of the supply chain and other novel procurement methods. These are fundamentally attempts to bring together the design and the construction activities by introducing those who do the building earlier into the design phase – into the front-end – or by improving the design–construction interface.

Kadefors (1999) believes that to integrate constructability/the contractor into the design phase is especially problematic as current contractual arrangements shape interests in such a way that problem solving and value management counteract each other. Traditionally contractors have not assisted in the design phase of buildings; nor have specialist subcontractors (e.g. mechanical or electrical) or large suppliers (e.g. steel merchants). However, Gunasekaran & Love (1998) argue that their contribution will be invaluable. These specialist organisations have specific knowledge concerning the capability of the life cycle of materials, the overall performance of a product and the programming of site operations. Khurana & Rosenthal (1998) believe that the real keys to success can be found by forging a holistic approach, particularly while two-thirds of the cost is committed during the design stage (Love *et al.*, 1998). The whole project is clearly planned out as far as possible enabling the production team to be appointed early in the process and, when required, to contribute to the design process and utilise resources effectively.

Research concludes that machinery, labour requirements and materials should be considered in the detailed or possibly the concept design phase (Riedel & Pawer, 1997). After all, the only way to improve a company's competitive position is by producing a design that is 'right first time' so that innovation and superiority can prevail, as changes at the design stage can be made efficiently and effectively thus reducing cost and time (Milburn, 1992; Riedel & Pawer, 1997; Khurana & Rosenthal, 1998). The ramifications of not considering the manufacturability/constructability of a product when it is being designed affect the quality, efficiency and speed of its development and it is imperative that the production aspect should be moved to the front-end (Ettlie & Stoll, 1990; Smith & Reinerstein, 1991; Cooper & Klienschmidt, 1994).

Cross-functional teams

Bringing the production team to the front-end of the process, as described in the previous section, invariably encourages greater interlinkages between the organisations in varying degrees throughout the project life cycle. This should be common practice. When construction problems arise the relevant organisations have to work together to determine appropriate concessions and compromises before solutions can be obtained (Alty, 1993). However, in practice this is rarely the case (Howell, 1999). Baxendale *et al.* (1996) suggest that the 'conventional delivery processes cannot effectively handle the interactions and complexities of a multi-disciplinary team approach', as current procurement methods exacerbate the division of tasks into functional disciplines that operate independently. Winch & Campagnac (1995) also share this common analysis that contractual arrangements only establish roles and responsibilities and do not give much insight into how the different actors are to interact. Moreover, the functional disciplines develop their own objectives, goals and value systems and become dedicated to the optimisation of their own function with little regard to or understanding of its effect on the performance of the project (Gunasekaran & Love, 1998). A construction problem has to be agreed upon by the various disciplines who each have their own perceptions and objectives so fundamentally the problem itself can possess multiple solutions (Li & Love, 1998). It is also common for many of the project participants to work on several projects simultaneously, leaving communication between the actors bilateral (Kornelius & Wamelink, 1998); skills are not harnessed and working relationships remain underdeveloped (Love *et al.*, 1998).

Conclusion

This chapter has set the scene for transferring product development principles to improve the design and construction process. The need for NPD and its generation of a series of process models was explained. The evolution of the construction process was described. It is clear that the *RIBA Plan of Work* is very much architect-led while the British Property Federation approach comes from the house-building sector; neither process is comprehensive. There is an obvious need for a more holistic and flexible process to enable the industry to develop a construction product for the twenty-first century especially through the use of increased front-end involvement and cross-functional teams. The implications from manufacturing are that better process definition is necessary; processes need to be managed and measured more effectively; and culture change, improved communication and common systems can be generated through the implementation of consistent process. Chapter 2 will present the tools and techniques that can be used to improve the design and construction process.

2

Techniques and Technologies for Managing the Product Development Process in Construction

'...defining and improving processes and finally applying technology as a tool to support these cultural and process improvements.'

(Egan, 1998)

This chapter introduces the techniques for process modelling and process improvement used in both manufacturing and the construction industry. The relationship between information technology (IT) and the design and construction process is defined and the technologies supporting each stage of the design and construction process are described through the use of an IT process map. This is followed by an account of developments in technology within the construction industry which are seen as crucial to the successful implementation of a common process and innovation (see Fig. 2.1).

Approaches to process modelling

An understanding of processes can be reached in different ways. The project process is often depicted/modelled to enhance team co-ordination and communication through simple mechanisms such as flow charts and Gantt charts (a flow chart that encompasses time). In order to model more complex scenarios of real-world phenomena techniques such as IDEF0 (integrated definition language) and analytical reductionism/process decomposition are commonly used (Koskela, 1992).

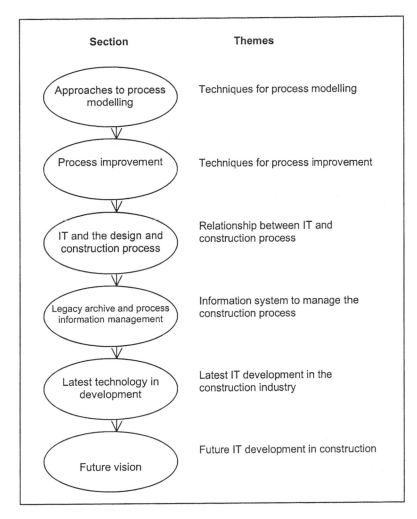

Fig. 2.1   Chapter map.

The integrated definition language (IDEF)

During the 1970s the US Air Force Program for Integrated Computer Aided Manufacturing (ICAM) sought to increase manufacturing productivity through the systematic application of computer technology. The ICAM program identified the need for better analysis and communication techniques for people involved in improving manufacturing productivity and thus developed a series of techniques known as the IDEF family (IDEF, 2002).

- IDEF0: used to produce a 'function model' – a structured representation of the functions, activities or processes within the modelled system or subject area.

- IDEF1: used to produce an 'information model' – represents the structure and semantics of information within the modelled system or subject area.
- IDEF2: used to produce a 'dynamics model' – represents the time-varying behavioural characteristics of the modelled system or subject area.

In 1983 the US Air Force Integrated Information Support System program enhanced the IDEF1 information modelling technique to form IDEF1X (IDEF1 Extended), a semantic data modelling technique. Currently IDEF0 and IDEF1X techniques are widely used in the government, industrial and commercial sectors supporting modelling efforts for a wide range of enterprises and application domains. In this chapter IDEF0 will be described as it most closely relates to the 'functional' new product development process.

IDEF0 for function modelling is an engineering technique for performing and managing needs analysis, benefits analysis, requirements definition, functional analysis, systems design, maintenance and baselines for continuous improvement (IDEF, 2002). IDEF0 models provide a 'blueprint' of functions and their interfaces that must be captured and understood in order to make systems engineering decisions that are logical, integrable and achievable to provide an approach to:

- Performing systems analysis and design at all levels for systems composed of people, machines, materials, computers and information of all varieties – the entire enterprise, a system or a subject area.
- Producing reference documentation concurrent with development to serve as a basis for integrating new systems or improving existing systems.
- Communicating among analysts, designers, users and managers.
- Allowing team consensus to be achieved by shared understanding.
- Managing large and complex projects using qualitative measures of progress.
- Providing a reference architecture for enterprise analysis, information engineering and resource management.

The modelling language itself makes explicit the purpose of a particular activity and is composed of a series of boxes and arrows. The boxes of the IDEF0 technique represent functions, defined as activities, processes or transformations. Each box should consist of a name and number inside the box boundaries: the name is of an active verb or verb phrase that describes the function; and the number inside the lower right corner is to identify the subject box in the associated supporting text.

The arrows in the diagram represent data or objects related to the functions and do not represent flow or sequence as in the traditional process flow chart model. They convey data or objects related to functions to be performed. The functions receiving data or objects are constrained by the data or objects made available. Each side of the function box has a standard meaning in terms of box/arrow relationships (Fig. 2.2). The side of the box with which an arrow interfaces reflects the arrow's role. Arrows entering the left side of the box are inputs; inputs are transformed or consumed by the function to produce outputs. Arrows entering the box on the top are controls; controls specify the conditions required for the function to produce correct outputs. Arrows leaving a box on the right side are outputs; the outputs are the data or objects produced by the function. Arrows connected to the bottom side of the box represent mechanisms; upward pointing arrows identify some of the means that support the execution of the function.

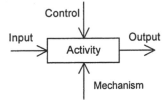

Fig. 2.2   The basic concept of the IDEF0 syntax (IDEF, 2002).

The functions in an IDEF0 diagram can be broken down or decomposed into more detailed diagrams until the subject is described at the level necessary to support the goals of a particular project (see Fig. 2.3). The top-level diagram in the model provides the most general or abstract description of the subject represented by the model. This diagram is followed by a series of child diagrams providing more detail about the subject. Each sub-function is modelled; on a given diagram, some of the functions, none of the functions or all of the functions may be decomposed individually by a box, with parent boxes detailed by child diagrams at the next lower level. All child diagrams must be within the scope of the top-level context diagram/parent box. In turn, each of these sub-functions may be decomposed, each creating another, lower-level child diagram.

Analytical reductionism/process decomposition

Process decomposition involves breaking the process down into levels of granularity as demonstrated in Fig. 2.4 with the lower-level sub-processes further defining their corresponding upper-level process. The level which differentiates a process from a procedure is, however, still a topic of discussion in the process management field.

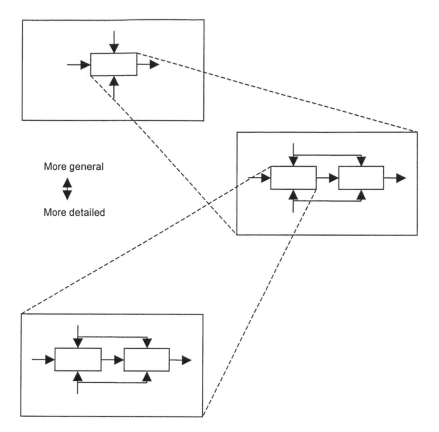

Fig. 2.3   IDEF decomposition structure.

The difference between a process and a procedure could be considered as follows.

A *process* (Koskela, 1992; Cooper, 1994; Vonderembse & White, 1996):

- Converts inputs into outputs.
- Creates a change of state by taking the input (e.g. material, information, people) and passing it through a sequence of stages during which input is transformed or its status changed to emerge as an output with different characteristics. Hence, processes act upon input and are dormant until the input is received. At each stage the transformation tasks may be procedural but may also, for example, be mechanical or chemical.
- Clarifies the interfaces of fragmented management hierarchies.
- Helps to increase visibility and understanding of the work to be done.
- Defines the business/project activities across functional boundaries.

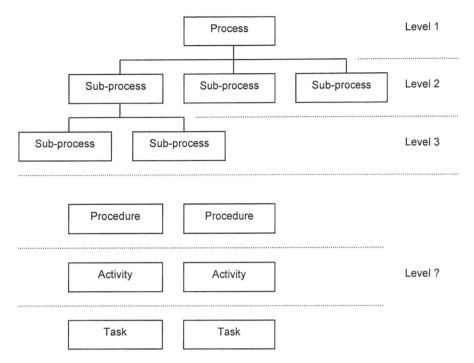

Fig. 2.4   Process levels.

A *procedure* (Lee *et al.*, 2000):

- Is a sequence of steps; it includes the preparation, conduct and completion of a task. Each step can be a sequence of activities and each activity a sequence of actions. The sequence of steps is critical to whether a statement or document is a procedure or something else.
- Is required when the task to be performed is complex or is routine and required to be performed consistently.
- Defines the rules that should be followed by an individual or group to carry out a specific task; their definition is usually rigid, leaving no opportunity for individual initiative.
- Supports the process.

Process improvement

The development of an improved process for design and construction will be illustrated in the next chapter. When a process has been defined or mapped it is then important to continue to improve the process. Process improvement is usually one of three activities:

- Management and continuous improvement of existing processes.

- Designing or redesigning of new processes.
- Concurrent engineering.

The first aims to optimise and continuously improve the on-going processes that are in operation within an organisation; the second endeavours to change the organisation's processes, perhaps somewhat radically. Both streams are explored below.

Continuous improvement (CI)

CI (sometimes referred to as continuous quality improvement, company-wide quality improvement, business improvement or process improvement) originated from the field of quality management. As its name suggests, it is concerned with the continuous improvement of an organisation's process (Kagioglou, 1999). McNair & Liebfried (1992) defines it as '...an incremental change process that focuses on performing existing tasks more effectively ... small improvements are made in the status quo as a result of on-going effects'. CI adopts the stance that creating a development process is never completed (Oakland, 1995) and improvements will only occur if attempts are made to learn from new information generated by the process itself rather than the product. This process is commonly associated with the plan-do-check-act cycle (PDCA; sometimes referred as the Demming wheel or Shewhart cycle; see Fig. 2.5; Oakland, 1995).

When continuous improvement is in place the PDCA cycle is repeated over and over again. Each phase of the cycle plays an important role in sustaining the ongoing improvement (Vonderembse & White, 1996):

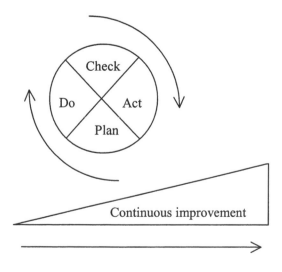

Fig. 2.5 The PDCA cycle (Oakland, 1995).

- *Plan*: identify problems or opportunities for improvement; develop a plan to make changes.
- *Do*: implement the plan, documenting any changes made.
- *Check*: analyse the revised processes to see if the goals have been achieved.
- *Act*: standardise, document and disseminate the results; in case of not achieving the goals, determine why not and proceed accordingly.

Business process re-engineering (BPR)

According to Davenport (1993) intense competition and other pressures on large organisations often overshadowed the quality initiatives that were introduced as part of continuous improvement programmes. Therefore a powerful new tool emerged to facilitate the fundamental redesign of work so that significant improvement could be achieved (Davenport, 1993), namely business process re-engineering (BPR; sometimes called process innovation, business process redesign, business engineering or process engineering). This is essentially a strategic business management theory and is now actively used in the construction industry. BPR differs fundamentally from CI in that it deals with breakthrough improvements as opposed to gradual changes (Wesner *et al.*, 1995). According to Ovenden (1994) improvements of 10% to 15% can be achieved in most companies by using CI. However, where quantum leaps in performance are required, where the old should be completely replaced by the new, re-engineering is the way forward and may well be the solution to improving construction. Furthermore, BPR processes appear to have a definitive start and finish whereas CI is a never-ending process (Zairi, 1991).

No two people have been more responsible for the re-engineering solution than Michael Hammer and James Champy (1997). The theory was well acclaimed after their book *Reengineering the Corporation: A Manifesto for Business Revolution* was published in 1993. They define re-engineering as:

> '... the fundamental rethinking and radical redesign of business processes ... to achieve dramatic improvements in critical contemporary measures of performance, such as cost, quality, service and speed'.

That is, not small incremental changes to the process while leaving the basic structures intact but rather abandoning long-established procedures, conventional wisdom and assumptions from the past to look afresh at the work required in creating a company's product or service to deliver direct value to the customer. An analysis of the definitions with respect to 'what', 'how' and the 'expectations' of the theory is given in Table 2.1 (adapted from Choi &Chan, 1997):

- *What*: the area of BPR concern.
- *How*: ways to deal with BPR as suggested in the definitions.
- *Expectations*: objectives expressed in the definitions.

Re-engineering is a top-down approach led by senior management and aimed at rapid and dramatic performance improvement (Ardhaldjian & Fahner, 1994). It views improvement from the process perspective rather than the functional or organisational stance (Klein, 1994b) and is intended to align the process with the strategic objectives and customers' needs. When a process is re-engineered jobs evolve from narrow and task-orientated to multi-dimensional. Moreover, managers stop acting like supervisors and behave more like coaches (Hammer & Champy, 1997). Workers focus more upon the customer's needs and less on their bosses'. Attitudes and values change in response to new incentives.

Although BPR has been a popular buzzword since it was widely pub-licised in the 1990s the verdict on its success remains uncertain (Choi & Chan, 1997). Hammer & Champy (1997) estimate that as many as 70% of the companies that try to re-engineer will fail but it is important to note here that this assertion was neither documented nor based on empirical evidence. Moreover, Champy reported in 1995 that re-engineering is in trouble in that even substantial re-engineering payoffs appear to have fallen well short of their potential:

> '. . . *Reengineering the Corporation* set big goals: 70% decreases in cycle time and 40% decreases in cost; 40% increases in customer satisfaction, quality and revenue; and 25% growth in market share … although the jury is still out on 71% of the ongoing North American re-engineering efforts in the sample … overall, the study shows participants failed to attain these benchmarks by as much as 30%.'

If the 'gurus' openly acclaim that there are risks attached to the process of re-engineering, why should we re-engineer? Chan & Peel (1998) conducted a survey of 37 companies from 17 different industries world-wide who had re-engineered and they concluded that about 73% of the sample experienced a significant reduction in cost while successfully fulfilling their customers' needs. Furthermore, 68% reported that re-engineering had helped them to achieve their objectives of improving organisational efficiency, making dramatic company-wide improvements and reducing throughput time. The study concludes that the efforts of re-engineering have more often than not paid off and that those companies that are willing to try can achieve sub-

Table 2.1 Definitions of BPR.

Authors	What	How	Expectations
Davenport *et al.* (1990)	Business process, IT, strategy, organisational structure	Analysis and design	Seeking for improvements
DeToro & McCabe (1997)	Cross-functional work teams	Examine, challenge and change work methods	Seeking for improvements
Dixon & Arnold (1994)	Key value business process	Radical or breakthrough changing	Dramatic orders of magnitude, distinguished from incremental improvements
Elzinga *et al.* (1995)	Systematic approaches to analyse, improve, control and manage processes	Process analysis	To improve the quality of products and services
Fiedler *et al.* (1994)	Business process information technology	Radical alteration, departing from existing practices, deliberating to plan	Bringing about remarkable improvements
Hammer & Champy (1997)	Business process performance	Fundamental rethinking, radical redesigning	Seeking for dramatic improvements
Kim (1994)	IT, business operations and organisation structures	Changing the way of doing business	Maximising the benefits of information technology
Manganelli & Klein (1994)	Strategic value-added business process, systems, policies and organisational structures	Rapid and radical redesigning	Optimising the work flow and productivity
Paper (1997)	Strategy-driven organisational approach	Redesigning critical business processes	Dramatic improvements in quality, cost, service and speed
Ryan (1994)	Company's market, customers, products, services, suppliers and competitors	Making fundamental changes	Seeking for improvements
Teng *et al.* (1994)	Existing business process	Critical analysis and radical redesigning	Breakthrough improvements in performance
Zairi (1997)	Manufacturing, marketing, communications	Analysis and design	To continually improve fundamental activities

stantial payoffs in several respects. However, there do appear to be a series of critical implementation strategies that increase the likelihood of success (Hammer & Champy, 1997; Chan & Peel, 1998; Revenaugh, 1994; Lee & Dale, 1998; Choi & Chan, 1997) and this can be attributed to the actual definition of the term (see Table 2.2). Hammer & Champy (1997) believe that the most egregious way to fail at re-engineering is by not re-engineering at all but rather conducting process changes and calling it re-engineering. The term has acquired a certain cachet and has been attached to many programmes that have nothing to do with radical process redesign. Bartram (1994) reports that some managers still admit that they still do not wholly grasp the concept. Furthermore, a survey showed that less than half of the 88% of executives who were re-engineering could successfully define BPR as process redesign (Klien, 1994b). The presiding factors are summarised in Table 2.2 (adapted from Choi & Chan, 1997).

Concurrent engineering

There are a number of well-proven techniques for supporting the concurrent engineering environment. Figure 2.6 (Browne & McMahon, 1993; Nevins & Whitney, 1989) illustrates the different tools and techniques as they are used at different stages of the NPD process. Many of these techniques are commonly used in medium to large organisations but their full benefits are not usually realised in small companies having few, if any, educated staff with exposure to these tools (Haynes & Frost, 1994). Figure 2.6 describes the tools and techniques used in concurrent engineering.

Quality function deployment (QFD) is a systematic approach for the design of new products or services based on close awareness of customer desires coupled with integration of corporate functional multi-disciplinary teams (Rosenthal & Tatikonda, 1992). It provides procedures to assist communications and structure decision-making by focusing efforts on identifying and providing the information needed for designing products and services (Griffin, 1992) thus bringing new products efficiently to market, increasing customer satisfaction (Liner, 1992) and decreasing product development time and cost (Eureka, 1987). QFD can be used in almost all the different stages of the NPD process, alone or in conjunction with other tools as shown in Fig. 2.6 (Browne & McMahon, 1993). It uses multi-disciplinary teams (Oakland, 1993) to translate customer requirements into the language of the engineer by explicitly linking the two kinds of information in a 'house of quality' (HOQ) (Griffin & Hauser, 1996). The HOQ is constructed using a series of matrices which match the customer requirements (*what* are the customer needs?) to the technical requirements (*how* can they be achieved?). In construction, where there is still relatively little

Table 2.2 Critical success and failure factors of BPR.

Category	Factors	Failure reasons	Success conditions
Definition	Concepts	Unclear definitions, too many terms and definitions, misusing the term	Understand and clarify the concepts
	Methodology	Lack of standard methodology	Select an appropriate approach, conduct change management, proper project management
	Expectation	Unrealistic expectations	Clear goal setting, expectation set on condition basis and review during the project
Human	Change resistance	Employees resist change, difficulty in culture shift, fear of downward decision-making authority	Establish communication channel, well-informed workers, provide training, care of those unable to adopt
	Top management commitment	Lack of top management commitment	Include senior executive in the steering committee, set strategic directions, having clear authority and project team motivation
	Workers involvement	Neglect line workers, depend only on outsiders	Balance between two parties, broad involvement of employees, comments for adjustments
Skill	Information technology	Over-reliance, mix up automation and BPR, lack of corporate information system	Maximising benefits of IT in BPR process, establish corporate overview at information flow, adequate budget
	Project duration	Delay of delivering the result	Proper project management, setting agreeable project duration, divide project into phases
	Scope and objective	Incorrect objective, wrong scope, wrong process selection	Prioritise objectives, establish business context, identify valuable processes
	Process re-conceptualisation	Limited by localised expertise, confusion of function with process	Need a shared vision for better communication
	Benefits recognition	Incapable of recognising the benefits of BPR	Encourage creativity, help by simulation

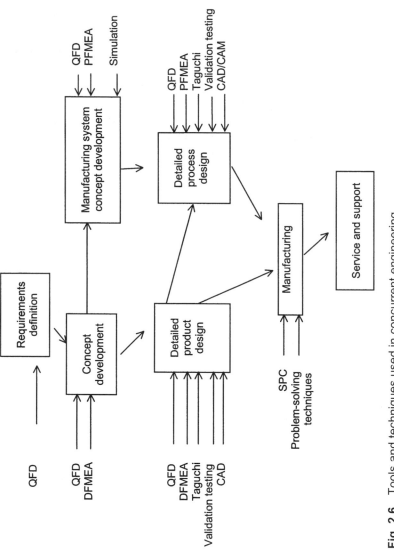

Fig. 2.6 Tools and techniques used in concurrent engineering.

focus on design, QFD would be most appropriate at an early stage of the process.

Failure mode and effects analysis (FMEA) is a 'summary of the engineer's thoughts as he/she designs a product or its components (design FMEA) or the associated manufacturing processes (process FMEA) and tools' (Ford Automotive Company, 1988). It is a systematic approach to eliminating possible and probable failure modes of the product and/or process, identifying their probability of occurrence and severity of failure and proposing corrective actions to be taken (Villemeur, 1992). It utilises cross-functional teams by including expertise from a variety of functional disciplines such as engineering and manufacturing.

Taguchi has produced a powerful paradigm consisting of three main elements for the general work plan of a manufacturing enterprise (Taguchi & Wu, 1980). The first, system design, generates the product specifications based on customer needs translated into design and manufacturing considerations; the second, parameter design, minimises variance in the product specifications; and the third, tolerance design, permits a bounded level of statistical variance rather than absolute precision (Cook, 1992).

Validation testing involves a number of test methods for verifying that a product design performs as expected, where statistical process control (SPC) is an effective way of measuring the performance of a process by gathering relevant information and using it to examine and identify possible corrective actions.

A number of other tools (philosophies) are also used to assist CE practices. They all aim at reducing development times and increasing the quality of the product. They include design for manufacturing assembly (DFMA), where manufacturing and assembly considerations are included in the design formation and selection process, achieving 'right first time'; and fault tree analysis (FTA) for decision making.

The basic techniques for developing and mapping processes and improving them have been described above. These can all be applied to construction which in turn can be supported by information and communication technologies.

IT and the design and construction process

Many studies have been conducted in the construction sector in recent years in order to investigate the relationships between IT and processes. Most of these studies have concentrated on IT capabilities and forecasting how IT will be used in the next 10 to 15 years (Brandon & Betts, 1995; IT2005, 1995; Building IT 2000, 1991; KPMG & CICA, 1993; Aouad *et al.*, 1997). These studies have predicted the types of technologies that will be used by the industry during that period. Construct IT (1995) produced an IT map that relates to the needs of construction processes without looking at the co-maturation of processes and IT.

However, the management of IT in construction has rarely been considered within a process context. Furthermore, both IT and process have frequently been treated as separate without any apparent links and/or interfaces. As we discussed in Chapter 2, there has been since the 1930s an apparent desire to change the construction cycle and several government and institutional reports have been produced to support this, including Simon (1944) and Banwell (1964), but none up to the Latham report (1994) has been significantly acted upon. During this time several protocols have been introduced including the *RIBA Plan of Work* (first published in 1964) and the British Property Federation manual (1983). These protocols have done little in considering IT as an integral part of the process. Hughes (1991) suggests that 'every project goes through similar steps in its evolution in terms of stages of work. The stages vary in their intensity or importance depending upon the project.' In the same way the IT elements remain the same but their use is dependent upon the project/process. The benefits of using IT in the construction industry have been illustrated by a large number of studies (see, for example, Betts, 1992; Brandon & Betts, 1995; Miyatake & Kangari, 1993; Teicholz & Fisher, 1994; Tucker *et al.*, 1994). However, IT has been introduced to the construction industry in a fundamental manner through its various professions.

Recent changes, both in the global market and in information and building technology, have begun to dictate a complete rethinking of the way we design and construct our buildings. As we have discussed, the construction industry, particularly in the UK, lags behind manufacturing in terms of both productivity and efficiency as there is no agreed procedural mechanism for doing the work due mainly to the fragmented nature of the industry and the operation of its activities, and the perceived uniqueness of construction projects. This also accounts in part for the slow uptake of new technology. To date IT has been used as a support tool rather than a driver in the design and construction process.

Traditional approach

The research by Childe *et al.* (1996) within the context of business process engineering (BPR) has shown that existing legacy IT systems are hindering the adoption of BPR principles by many large organisations. In a survey of 34 companies (none of which was from the construction sector) it has been demonstrated that these existing IT systems are blockers rather than enablers of process improvement. The construction industry is not susceptible to the same problems as many of the existing systems have been acquired on a relatively smaller scale; thus upgrading or even replacing them will not be a difficult task in broad financial terms but rather will offer the opportunity to ensure compatibility and 'fitness for purpose' from the onset.

The main problem in construction is that most of the IT systems have been purchased in the past because of operational rather than strategic/business requirements. These systems have failed many construction firms, leading to some suspicion of what IT can deliver to the sector. In order to rectify this Alshawi & Aouad (1995) proposed a framework that addresses the significance of merging information systems (IS) and their associated IT strategies with business objectives. However, this work has failed to look at the co-maturation of processes and IT. A new process/business view to develop IT solutions through an IT map was developed by the authors.

The IT map (Fig. 2.7) identifies technologies that enable specific processes within the design and construction cycle to have better performance through a co-maturation model. The IT solutions are classified under major headings including communication, visualisation, integration and intelligence. These technologies are addressed in terms of their maturity in relation to the processes they are trying to support.

IT process map

The IT map presents a vision for the future regarding IT in construction. It is acknowledged that the potential benefits that will come with the improved process can only be realised with significant IT support. However, the IT will only achieve profound change if its introduction and use is linked to changes in the overall conduct of the design and construction process (Aouad *et al.*, 1997).

The IT map is developed around the structure of the Process Protocol Map (see Chapter 3). It includes technologies specified by industry and academia at the initial stage (pre-project phase) where simulation, 'what-if?' and economic appraisal tools are most appropriate. Potential project scenarios could be generated from an archived library of previous projects. AI (artificial intelligence) techniques including case-based reasoning, neural networks, fuzzy logic, genetic algorithms and KBS (knowledge-based systems) may also be appropriate to enhance creativity in the initial production of the design. Economic appraisal cost-planning tools can be applied to compare different alternatives and perform feasibility analysis at this stage of the process.

In the pre-construction stage the client's needs are developed into a

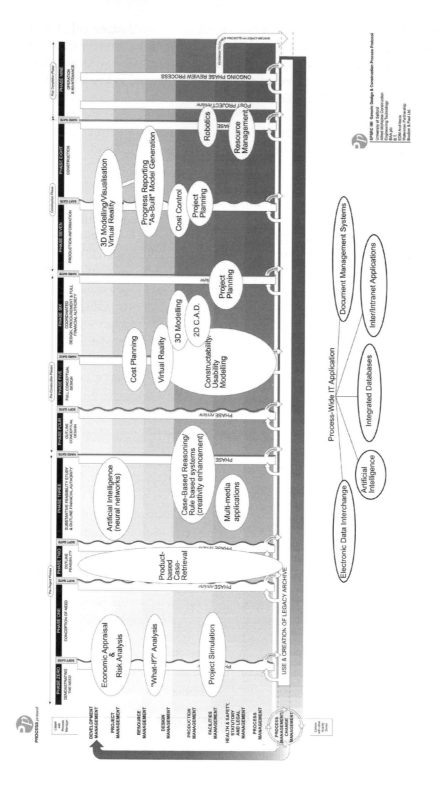

Fig. 2.7 IT map.

design solution. This requires the use of AI, CAD (computer-aided design) and VR (virtual reality) tools. The key themes of visualisation and intelligence are dominant at this stage. Visualisation supports the simulation of information within the various phases of the process. Technologies which can be used include VR, CAD and simulation tools. Artificial intelligence helps in establishing decision support systems which may manage to automate some of the phases of the process. These include knowledge-based systems, neural networks, case-based reasoning and information management systems.

The stages of construction and post-construction involve the use of legacy applications such as planning and estimating. Integration and communication are the key themes at this stage. Common standards, STEP (Standard for Exchange of Product model data), IFC (Industry Foundation Classes) and tools such as email and web applications are often deployed at this stage of the process. Finally integration technology allows sharing and exchanging of the project information across the project teams. Major research in this area includes; ICON (Aouad *et al.*, 1994), OSCON (Aouad *et al.*, 1996), SPACE (Alshawi, 1996), COMMIT (Rezgui *et al.*, 1996), IDAC-2 (Powell, 1995), COMBINE (Augenbroe, 1994; Dubois *et al.*, 1995), ATLAS (ATLAS, 1992), MOB (MOB, 1994), COMBI (Ammerman, 1994), RATAS (Bjork, 1989), IRMA (Luiten *et al.*, 1993), Fenves *et al.* (1990), Froese & Paulson (1994), Howard (1992), Kartam (1994). All of this research has recommended improved systems as the future for design and construction.

A driver or an enabler?

Ideally the process, IT, people, culture and customer issues need to be considered and developed together to produce a comprehensive model. Such a model would identify the enablers for process execution as IT, people and culture. For example the type of IT will probably depend on the people who use it and the extent of its use will depend on the culture of the organisation and/or industry. An 'open' culture will utilise the communication capabilities of an IT tool to ensure visibility of project elements and deliverables to all parties in a project. The effective use and co-ordination of IT, people and culture interfaces should optimise the process performance which leads to eventual customer satisfaction.

There are six critical dimensions of IT involved:

(1) Simulation (e.g. 'what if', project simulation, economic appraisal).
(2) Integration (e.g. integrated databases).
(3) Communication (e.g. EDI (electronic data interchange), Internet).

(4) Intelligence (e.g. artificial intelligence, KBS, neural networks, case-based reasoning).
(5) Visualisation (e.g. VR, 3D).
(6) IT support (e.g. CAD, project planning, cost control).

The common problems associated with IT are related to its uptake, which has been apparently relatively unco-ordinated, and the fact that its strategic application appears to have been determined by availability of IT rather than its suitability. The unsuitability of IT systems causes disfunctionality in the process infrastructures which they are expected to support. It is evident that the uptake of IT systems by the industry has been broadly technology-led with the industry using basic communication tools in a widespread (but not comprehensive) manner; and that the application of particular industry-specific tools is more localised, probably because of communication problems. In part this is due to a lack of understanding of the way in which organisations and their operational and managerial processes operate compounded by a lack of appreciation of how information technology supports them. At a more sophisticated level of analysis, the organisational capability and maturity of a company (or industry) are related to a number of issues including the role of process management and information systems in their maturation.

The technological management of IT within the construction industry has been given little attention in the past. As a result a number of IT solutions have been developed to act as drivers in the design and construction process. IT-enabling solutions will play a major role in achieving significant improvements in a traditionally fragmented design and construction process. This can be achieved by viewing a process in two dimensions:

(1) There must be process and IT alignment. The IT process map presented here illustrates how IT could operate within a process framework. This requires an agreement on the actual design and construction process phases, structure and management. In such a way the process becomes the driver and IT the enabler.
(2) The phenomenon of co-maturation of IT and processes needs to be considered. IT can only be effective if it is based on synchronised process development. For example, the full benefits of an optimised process cannot be realised when the IT development is still at the *ad hoc* stages, and *vice versa*. Within this framework, it is anticipated that construction firms will move away from traditional *ad hoc* IT investments and move towards well-planned strategies. By doing so large as well as small organisations will be able to identify opportunities for IT investments, evaluate their existing systems, identify the rate at which new IT applications are adopted and work out the level of impact of IT

on their firms. Essentially it is suggested that the coupling of processes, IT, people and culture will provide customer/client satisfaction accompanied by many obvious benefits.

Legacy archive and process information management

Construction projects by their very nature produce a great deal of information regarding the design, the project management and the construction attributes. Most participants then move on to the next project and little learning takes place either throughout the project or between projects. A *legacy archive* was a proposition identified by the authors through their work on process during the 1990s. This is a 'live' communications link which not only facilitates co-ordination on a project-based environment but also offers the possibility of learning from previous project successes and/or failures (Kagioglou *et al.*, 1998a). Potentially an IT solution should be in place for the project team and the organisation(s) as a whole. What needs to be considered is how the information or project knowledge can be reused in the other projects, how the knowledge can be captured and what technology can be used to realise it. These are issues of knowledge management; that is, management of the project knowledge. However, most of the common knowledge management systems focus on processing and sharing information and are not designed to accommodate the specific process framework. To achieve effective project information management project information has to be managed in a well-accepted information framework. The legacy archive was specified by the Process Protocol project (see Chapter 3).

The legacy archive proposal is for a process management tool which is a knowledge-based project information management system that integrates the process model as its core information framework. It should provide:

- Knowledge-capture functionalities such as document/drawing publishing to record project activities and archive project documentation based on the process framework created by the *process creation tool*.
- Knowledge development functionalities using data mining or OLAP (on-line analytical processing) techniques to analyse the project information and identify the information pattern, potential conflict such as potential resource issues on site, optimised construction programme and construction process simulation for possible crash detection.
- Knowledge-sharing functionalities such as messaging service, email notification and document sharing.
- Knowledge-utilisation functionalities, web-based interface, a personalised project information page for each user and a fast search tool for document or information retrieval.

In this environment teams can reduce costs and save time as they gather and disseminate information throughout the project life cycle. Furthermore, the integrated project process map will become the route map to help and guide the project management team to monitor and track project progress and documents. The centralised project information can be reused in a future project where the knowledge captured and stored can be used to inform decision-making and value engineering.

Latest technology development

Project extranet

Recently the Web has facilitated approaches to achieving a legacy archive. Web-based project extranet is one of the solutions for improving co-ordination, collaboration and communication across construction projects which has been growing rapidly since 2001. Project extranets have been around for some time but were expensive and difficult to use and thus were only used in some major projects. However, with the wide availability of the Internet and the development of the technology, modern project extranets have become cheap and user-friendly business tools. Web-based project extranet first started in the USA in 1994 and brought the construction industry into the e-business age. It enables the members of the project team to communicate with each other and share the project information via the Internet in a controlled and secure way.

The main function of project extranet is to share project documents. The current systems are now not only allowing the sharing of documents through the network but also enabling users to view the maximum number of CAD files without installing any extra software. Moreover, users are able to mark-up (redline, comment) revisions which then become a part of the original document.

As project extranet is a restricted network for the project team, every user is identified by user ID and password. It is possible to automatically track and log the activities of individual users and identify who made which comment or who viewed or modified a particular file. It is also possible to introduce the project hierarchy and assign different access levels to ensure that information is seen by the right group of users.

Another important advantage of project extranet is that all the project participants have access to the most up-to-date versions of project information. This will in theory significantly reduce the expensive mistakes caused by someone working on out-of-date information which are quite common in the construction industry. Furthermore, the cost of sending and printing the project documentation will be reduced since most of the documents will be exchanged electronically.

Building information model

The concept of a building model was introduced in the 1970s. Much of the research in this field has been undertaken by academics, for example the OSCON (Aouad *et al.*, 1996) and Gallicon (Aouad *et al.*, 1996) projects at the University of Salford. Since 2003 leading CAD vendors, such as Autodesk, Bentley and Graphisoft, have been heavily promoting the concept as the building information model (BIM). The BIM is a digital data model of building design information which may also contain information about the building's construction, management, operations and maintenance. Unlike traditional 2D CAD systems in which the building design is represented in multiple drawing files made up of lines, arcs and circles, the BIM is a single information model constructed with intelligent 'objects' which represent building elements such as walls, slabs, roofs, doors and windows. If a window is moved in the elevation the corresponding plan or section will be updated automatically to show the window's new location. If the design parameters of the window are changed in the window schedule those changes are automatically reflected in the drawing views as well.

Various technologies and researches have demonstrated the benefits of the building information model concept (Lee *et al.*, 2003). However, these technologies are based on different standards that are not compatible with each other. An open and neutral data format is required to ensure data compatibility across the different applications. IFC (Industry Foundation Classes), developed by IAI (International Alliance for Interoperability), provide such capabilities.

Future vision

Recent years have seen a major change in the approach to construction innovation and research. There has been a huge concentration, from both the academic and the industrial communities, on the development of a single building/product model and/or on the expansion of 3D CAD modelling with other design attributes (such as time and cost) thus forming an integrated model that is shared by all key participants in the project process. The information in the model is linked so that when design information is changed, for example, the cost of the project will also change to reflect the new design. The integrated model will be a multi-dimensional computer model that will portray and visually project the entire design and construction process enabling users to 'see' and simulate the whole-life of the project. This will help to improve the decision-making process and construction performance by enabling true 'what-if' analysis to be performed to demonstrate the real cost in terms of the variables of the design issues. It will therefore be possible to clearly envisage the trade-offs between the parameters, such as:

- Predicting and planning the construction process.
- Determining cost options.
- Maximising sustainability.
- Investigating energy requirements.
- Examining people's accessibility.
- Determining maintenance needs.
- Incorporating crime deterrent features.
- Examining the building's acoustics.

This approach to the future, combining multi-dimensional versions of the building aligned against the design and construction process, is under development in research establishments and organisations worldwide. The technology is available, IT is an enabler and we must establish strategies to select such technologies. However, before IT can be used for the benefit of the industry there is still a lot to do in understanding and developing the processes and training the people in the industry to work with them. Chapter 3 illustrates through a case study how a generic process was developed with industry.

3

Case Study: the Generic Design and Construction Process Protocol

'The starting point for improving construction is to change the way of thinking, rather than seeking isolated solutions to the various problems at hand'.

(Koskela, 1992)

'The task of creating a development process is never completed. Once we have designed the process we need to change it to keep up with changes in the environment. This process improvement will only occur if we attempt to learn from new information generated by the development process itself. Importantly, this is not information about the product, but rather information about the process.'

(Reinertsen, 1997)

This chapter describes the development of the Process Protocol, a generic design and construction process for use by all sectors of the construction industry. It describes the construction and manufacturing experiences that were to formulate the principles and eventual structure and form of the Process Protocol. It then goes on to describe the Process Protocol itself.

The Process Protocol was developed between 1995 and 2000. The first

Unless otherwise stated, sources and analysis of factors are derived from industrial feedback gathered from interviews, questionnaires and structured/semi-structured workshop discussions involving the industrial partners to the project: BAA, BT, Alfred McAlpine (Special Projects), Watermans Partnership, Capita, Boulton and Paul, and Advanced Visual Technology.

high-level process was developed by a research team at the University of Salford in the UK in conjunction with nine collaborating companies representing the spectrum of industry participants. This work was built upon over the next three years by the same group with additional expertise from Loughborough University. Throughout its development the Protocol was tested by a whole host of industry collaborators.

In conjunction with the review of contemporary industry literature and published research findings, the research team had the benefit of access to workshops, questionnaires, interviews and other contributions from professionals representing the industrial partners in the project team. These combined inputs were available over the project duration. Perspectives were provided from two major clients both of whom were exploring the development in their own client-led construction processes. Design and structural engineering companies, construction/project management specialists, construction information technologists and a specialist product supplier were also party to the development of the Process Protocol. Their input informed and steered the development of the Process Protocol research, enabling the research team to correlate the identified needs of the construction industry for change with a combination of research and practitioner-based experiences.

The description of the development of the Protocol which follows is therefore a combination of referenced research findings and experiential input from the industrial partners.

What do we mean by 'process'?

The first issue in conceptualising a new process was to define the scope and perspective of the term process. A review of the academic construction literature and references in other construction industry publications indicated that the definition and perceived scope of the term was highly variable particularly within the construction field. It was also distinctly different from the definitions assumed in the various manufacturing research fields consulted. In particular, there were significant differences in process perspectives between construction and what appears to be the most comparable manufacturing process, new product development (see Chapter 1). This was amplified through discussion with the industry partners and it took several months and meetings to agree a definition of process. Indeed the research team found very little common understanding of the term in the industry, despite its continual use in the management arena with reference to concepts such as business process engineering.

Conceptualising a model

Given the apparent lack of commonality in the contemporary under-standing of the design and construction process, one of the first activities the research team did was to attempt to produce a model of the process which could be debated and subsequently refined towards a generic representation.

A number of modelling/systems analysis techniques such as procedure charting, data flow diagrams (DFD), system flow charting, HIPO (hierarchy + input-process-output) and data modelling were considered. The IDEF0 (integration definition language 0 for function modelling) process modelling technique was adopted, initially, as the most appropriate means of modelling the process. In developing a process model using the IDEF0 technique an initial step is the establishment of the activities that will comprise the model. In preliminary workshop sessions with the industrial partners these activities were presented for discussion in the form of an activity hierarchy (similar to an organisational chart). Initial reactions to this were poor, principally because such an approach did not facilitate communication of the process either quickly or clearly.

Although IDEF0 provided detailed semantic information, it tended to become more complex as more sub-processes were added into the model, and this resulted in difficulties in communicating the model. Moreover, it was found that the partners, at this stage, also preferred to concentrate on the general principles of the process rather than the detail of the activities involved.

This preference for generic principles was found to have a certain congruence with other models of manufacturing processes. In such models the graphical representation of the process conveys its inherent principles. As Rosenau (1996) notes, such process models are 'an effective way to show how a process works'. As a definition:

> 'A process map consists of an x and a y axis, which show process sequence (or time) and process participants, respectively. The horizontal x axis illustrates time in process and the individual process activities or gates. The y axis shows the departments or functions participating in the process...'

Beyond this convention there appears to be little formality in the method used to represent a process. It was decided therefore to use a more simplistic graphical representation of the process. Through interaction with the partners this relative informality of the modelling process enhanced the contributions of the project's partner representatives. Once an outline model was drafted, through several workshop sessions, the model was revised and deliberated by the partner representatives. As Rosenau (1996) argues, this 'participative' approach to design makes any new process easier to accept and use. In an industry with a 'need for change' such an approach is the most appropriate as it allows the participants to have ownership of the model they are creating. Thus through a process of gradual refinement progress was made towards an agreed version that was based on an informal information (activity) flow-type modelling technique.

Drivers behind the process

As the modelling developed, the workshops and interviews identified a number of underlying factors which provided a direction for the development. Contributions to these came from both the construction industry and manufacturing industry.

Construction and 'process'

Construction industry perspectives on operational and managerial 'process'

Some sectors of the construction industry and the construction research field clearly perceive solutions to process problems only in terms of the scope for construction operation/activity-driven improvements (Skibniewski & Molinski 1989; Lutz & Hijabi, 1993). Conventionally cited construction industry distinctions, such as the scale and unique nature of the construction product and the normal construction of buildings in their place of final use, compounded by an industry focus on *construction* rather than on the *assembly of components* on site, are usually considered as major inhibitors of adopting manufacturing approaches to process. Therefore the argument is that process improvement must be focused on the construction process rather than on design and construction. There also appears to be a focus on consistently improving through simplifying (Burbidge, 1990; Plossl, 1991).

In addition many of the production process improvements have been driven by technological innovation or operational standardisation (Lutz & Hijabi, 1993). However, they were by their nature diverse and uncoordinated – hence a strategic management process was required to co-ordinate their potential benefits and realise their potential. The Process Protocol was therefore not targeted purely at the production processes. Notwithstanding this, ever-increasing prefabrication and standardisation of components and sub-assemblies for building production is creating an emergent standardisation of elements or modules of production which a consistent managerial-level process would probably help stimulate and support.

Another problem in changing the industry was the attitude towards process in construction. Even in the post-Latham era of change, the construction industry focuses on the goal of a changed industry rather than the journey and process of change necessary to approach and refine that goal. The difficulties associated with reconceptualising the manufacturing process were not dissimilar (Koskela, 1992). Plossl (1991), reflecting on the comparison of manufacturing and construction processes, commented in 1991 that:

'the characterisation of manufacturing could as well describe construction ... the consensus of practically all people in manufacturing until recently was that the problems experienced daily were inevitable and that it was necessary to learn to live with them. The real heroes were those individuals who could solve problems shortly after they arose, regardless of how they solved them.'

The process would therefore need to address these issues and embrace more than mere production. It would also have to consider how to use the output as a way of changing attitudes, many of which were evident in some of the partner companies.

The need for process to link design and construction phases

Problems of component and product variability arising at the interface between operational activities are also commonly interpreted by the construction sector as being influenced by design decisions. Any resultant problems are commonly seen as an uncontrollable external factor in operations which occurs as a consequence of a lack of common understanding between the design perspective and the assembly practicalities and is compounded by poor communication.

The historical linear, sequential relationship between design and construction activities is a frequently cited contributor to buildability problems due to a lack of common perception/understanding and/or communication between designer and constructor. It was a clear priority from the outset of this project that the design and construction phases required to be better integrated – both in the context of a common managerial approach and in enhanced communication (including its content). Since this could not be achieved at the (post-design) operational level alone it would have to be achieved using a strategic management level that spanned the design and construction phases. This also began to highlight the need to consider revising the duration and entry point of professional involvement in the process in relation to the life cycle of the product. Specifically, it raised the issue of earlier co-operative involvement of many professional roles in the design and/or pre-project phases.

Manufacturing and 'process'

In order to set the development in context the research team undertook research in the manufacturing sector which was then presented to the industry partners. This helped to illustrate how and where we could transfer good practice between sectors.

A review of the manufacturing research field (see Chapters 1 and 2), and in particular the field of new product development (NPD), indicated distinct differences between manufacturing and construction in the involvement of designer and assembler in the NPD process and corresponding differences in perception of the terms of reference for process and process involvement. In new product development it appeared to be more conventional for product developers to be involved in the early stage of a client's exploration of a new product (Cohodas, 1988). This is sometimes referred to as the requirements capture phase which is characterised by a consultative involvement of the potential product developer(s) before the decision to proceed with the product is finalised. A further issue arising from the review of NPD processes was the use of an explicit stage-gate process approach whereby decisions on the project progression were scheduled for a number of prescribable decision points. Scope for the concurrent process planning and professional input was included in such process philosophies which in turn led to scope for the projects to be systematically conceived, developed and revised while retaining the option to vary or cease the process within the prescribed process management protocol (Cooper, 1994) (see Chapter 1). In NPD this was reported as both allowing flexibility of product and operational processes and supporting the innovation step within a generic set of strategic management principles (Cooper, 1994). In addition the potential introduction of 'fuzzy' gates within the process could further facilitate reduced development times and ease the natural progression of a project's life cycle (Cooper, 1994). This approach became a key parameter for the Process Protocol.

The NPD process approach also commonly involves the product developer in the review of products after their completion, which allows subsequent versions to be improved. It was therefore clear that for some sectors of the manufacturing industry involvement was more intense, interactive and relevant for a larger proportion of the product life cycle. Translated to construction, this appeared to offer extra scope for a pre- and post-production consultancy and evaluation role and also to allow for the construction industry to generate and use feedback in improving the product and their own production process. Client expectations of the manufacturing sector clearly embraced early consultation and the use of feedback as part of a commitment to quality improvement, facilitated by long-term co-operative arrangements or strategic partnering and an appropriate fee structure for pre-product consultation. The emerging practice in world-class manufacturing of reviewing the *process capability* of bidders as well as the product quality is reported as helping those clients who themselves are capable of looking beyond the product (Foxley, 1996).

These concepts had a profound impact on the conceptual design of the Process Protocol.

While changing the extent and duration of involvement of the construction industry in the life cycle of their products would require facilitating changes in contractual arrangements, there appeared to be clear potential benefits for the client *and* the industry – in terms of diversified and increased demand for their professional knowledge and the chance to use this as a mechanism for improving products and processes. Coupled with the raising of client expectations by their exposure to manufacturing service and processes, this is probably going to be a growing and significant pressure on the construction industry to change its service. Research undertaken on partnering by investigators working on this project indicated that a partnered arrangement may assist in facilitating such developments. The terms of reference for 'process' were therefore extended to cover the early stages of the life cycle of products and the latter stages of their use.

The terms of reference for designing the Process Protocol

The construction industry and research field foci prioritised by the industrial partners were the pre-completion processes of designing and constructing buildings and not, for instance, the process analysis of facilities management of completed buildings. One reason for this was that apart from PFI (Private Finance Initiative) and BOT/ BOOT (build-own-operate-transfer) type projects, which extend the involvement of the construction industry beyond completion of the building, the majority of potential industry beneficiaries of a generic Process Protocol would be contributing to the pre-completion phases of the building life cycle. However, with public-private partnerships, the Public Finance Initiative and similar approaches on the increase, further work is essential and underway to link the Process Protocol into overarching business processes.

With the benefits of steering input from the industrial partners, the development focus of the Process Protocol was therefore deliberately targeted on the design and construction phases. This was informed by a general recognition that poor process control and performance in the early stages of design and construction led to compounded problems for the latter stages of the process and therefore the most achievable and effective opportunity for improving overall process management and performance lay in the early stages of the process.

Strategic level of the Process Protocol

It was also appreciated that the fragmentary nature of the construction industry had been and would otherwise continue to be a major obstacle to achieving co-ordinated control of the design and construction process. As discussed earlier, it was felt that a process standardisation focusing on operational activities in the construction phase was insufficiently strategic to be effective; also it could clearly not support the co-ordination of the entire process which included design and use as well as assembly. Furthermore the incremental increase in standardisation of components and building sub-assemblies was already producing convergences in site operations and forms which would develop best without attempts at generic orchestration. A management-level protocol may facilitate and stimulate increases in operational-level consistency and this was considered to be a possible side-benefit. The subsequent development of tools and techniques for implementing the Process Protocol and facilitating industry change would clearly have to focus on these issues but they were outside the scope of the funded project brief.

Generic nature of the Process Protocol

The industry partners were clear that the protocol must be designed as a prototype or generic tool which could be adapted and applied irrespective of the variability in particular project details; in other words it must be a generic strategic management protocol. The selection of process factors to address was related to the operation of the various professional functions, their co-ordination and their relationship to the overall process of producing a building that met the client's needs and expectations.

Factors identified with problems in the design and construction process

As discussed above, the key priorities identified by the industrial partners for developing the Process Protocol were the need for managerial-level change and, in particular, the need to orchestrate existing management-level activities within design and construction into a *coherent* and *repeatable* process capable of *systematic* and *sustainable improvement* (Hinks *et al.*, 1997).

In addition, a number of factors identified from previous research as problems or contributory factors in poor performance were analysed and used to design the improvement parameters for the process protocol. These included the following.[1]

Process problem factors identified for the pre-project phases of the design and construction process

Pre-project was defined as comprising those phases which occur before the decision is made by the client to proceed with a construction solution to their business need. These phases correspond to recognised phases in NPD (Clark & Wheelwright, 1993) but conventionally occur prior to construction industry involvement. NPD pre-project phases include requirements capture and outline and detailed feasibility studies produced for the purposes of selecting a preferred business option, of which construction may be one of several potential options.

- Demonstrating the need for the project

There was a clear identification in the research that the early stages of the life of a project were poorly handled by the construction industry – a problem which was in part fuelled by the contract-led restrictions on involvement of the various parties in the very early stages of a project's life and compounded by the lack of a convention of being involved in a pre-project on a consultative basis.

- Consultative involvement

The scope for consultative involvement of producer/developers which occurs in the distinctive NPD sector requirements capture phase was not generally available to the construction industry or its clients which also

---

[1] Unless otherwise stated, sources and analysis of factors are derived from industrial feedback gathered from interviews, questionnaires, and structured/semi-structured workshop discussions, involving the industrial partners to the project: BAA, BT and Alfred McAlpine Special. Projects, Watermans, Capita, Boulton and Paul, Advanced Visual Technology.

meant that the opportunity for the design and construction professionals to provide a professional service was not generally occurring. In manufacturing NPD sectors the arrangements can operate on a retention and/or consultative fee basis, frequently facilitated through an implicit or explicit partnering arrangement. It is from the requirements capture phase onwards that the opportunities for partnering relationships commence in NPD which in turn facilitate mutual strategic benefits for the partners. While not without example in the construction industry, this is still a rarity even in emerging partnered relationships.

- ## Partnering opportunities in pre-project phases

Partnering will not be able to bring about the necessary change in process unless it provides a strategic relationship that stimulates and supports pre-project consultation. Without a change in contractual and/or process conventions project-specific partnering is unlikely to be able to provide this potential benefit since the involvement of the partner may not occur until at or near the conventional industry entry stage when a client has already committed themselves to construct .

The process of pre-project consultation

In projects where there is not an alternative provision such as a consultative retention/strategic partnering relationship the difficulty may be for the client to know:

(a) Who to approach for advice on a consultative basis.
(b) How to manage the process of taking consultation advice into their decision-making process across what may be several internal phases of deciding about a project.

In NPD processes the decision-making phases which can occur prior to committing to construct as the business option can include:

(a) Demonstrating the need to internal assessors.
(b) The conceptualisation of the demonstrated need to define the scope and options for a potential project (of which construction may be but one competing option).
(c) Conducting various levels of feasibility study for the client body to select the preferred options.

Only at this stage, if construction is the chosen option to meet their business needs, may the construction industry be approached under the contemporary conventions. This arrangement leaves the industry knowledge under-utilised and frequently responding to ideas which the client has committed themselves to on the basis of little or no professional input. There may also be a body of opportunities to construct which are not reaching the industry because the client is unable to get access to the advice required to assess feasibility.

- ## Client briefing: interface, access and communication

The problem factors associated with these phenomena in published research findings included poor client briefing (Jamieson, 1997) and unclear or missing information (BRE, 1995) which affected subsequent phases of the design and/or construction process. This is probably compounded by difficulties in accessing the client brief (Jamieson, 1997) which, with an earlier and more meaningful involvement of the key design and production parties, may lead to mutual product and process benefits. There were also clear problems recurrently identified with the interface between the client and the industry, usually between client and designer, and associated with the initial briefing exercise. The need for improved communication is a frequent factor that the industry is alerted to but the issue is deeper than merely communicating – it also includes what is being communicated. These problems appeared to be perpetuated by a lack of involvement of appropriate expertise in the very early phases of product conception that could lead to subsequent poor co-ordination in the design or project planning phases (Jamieson, 1997).

- ## Existing design and construction 'processes'

Problems identified with the inconsistency, linearity and sequential nature of existing design and construction 'processes' (Jamieson, 1997; Cooper, 1993) have been linked with the difficulties in producing and using construction production information (see also Mokhtaar *et al.*, 1995). The problems with the management and co-ordination of production information have been associated with the process of their production. These problems are frequently compounded by an excessive but unco-ordinated and unstructured bureaucracy in client bodies (Assaf *et al.*, 1995). Productivity lags behind manufacturing (Koskela, 1992).

- Requirement for industry-wide coordinated process improvement

It is essential that the parties involved in project teams are operating to a consistent process, which requires broadly similar process capabilities of the various team members. This will mean that industry-wide process improvement is required for the design and construction process to achieve repeatability and hence manageability (Hinks *et al.*, 1997). This means linking the professional capabilities in the pre-construction phases as well as the production phase parties and linking capability and process involvement across the phases. The alternative could be a greater level of stratification and fragmentation within the industry which would be self-defeating since the problems of process depend on all parties for their co-ordinated resolution.

- Assessing process performance

The performance of a structured process could be assessed by the stake-holders using a phase review board system similar to that employed in NPD. Controlling the process so that the changes are restricted to the product and not the process would allow consistency of management control to be retained. Use of a consistent and controlled process would enhance capability to accommodate product changes. The need for generic process models was identified by the industrial partners (Whitelaw, 1996) and in some academic literature (Jamieson, 1997; Whitelaw, 1996; Aouad *et al.*, 1994; PISTEP, 1994). There was also a call for integrated processes in construction (Sanvido, 1988) and data management protocols (Jamieson, 1997).

- Barriers to learning

A product-orientated emphasis (rather than a process-orientated emphasis) focuses on the differences between products and processes operating in other sectors. One-of-a-kind products do tend to limit the feedback potential (Koskela, 1992) and focus reviews on product feedback and not process feedback. By focusing on the uniqueness of the products, rather than the commonality of the process for the management of their production the construction industry continues to put its efforts into solving and resolving individual product-focused problems without creating the managerial systems which could help avoid, overcome and/or learn from (product and process) solutions.

Many construction projects originate design/construction information and data which may have been gathered from previous projects and improved following feedback if there were an appropriate mechanism for doing so. The need for designers to interface with previous and ongoing design and production data for the specific project (Mokhtar *et al.*, 1995) and from previous projects is a potential aim for construction IT especially where as-built CAD drawings are produced from relatively conventional (repeated) building forms. At the centre of this is the complexity of information flows (Jamieson, 1997). Problems with compatibility of the IT have tended to obscure problems with incompatible processes and professional conventions/boundaries which will also probably restrict integration. In the 'non-repetitive, complex and dynamic' construction industry computer systems 'have been found wanting' (O'Brien, 1997) (see Chapter 2).

Risk assessment

This was identified by the industrial partners as one of the most critical yet unco-ordinated activities for modern project/process management. A lack of follow-on of risk assessment and decision data analysis between phases of the process was also identified as compounding the unsystematic risk assessment within phases (Jamieson, 1997). Much of the risk assessment being undertaken also appears to be being undertaken by individuals rather than all the stakeholders in the project in a co-ordinated and consistent manner (Shahat *et al.*, 1995). A clear protocol for the early addressing of risk and risk allocation by the stakeholders in the process was considered essential. A lack of review of risk assessment meant that there was no systematic feedback to accommodate improvements in future risk assessment. This produces a consequent compounding of errors throughout the process (Jamieson, 1997). Risk assessment was also felt to be an essential early process phase activity which construction professionals could assist clients in undertaking pre-project-phase feasibility analysis.

There is also a separate problem of the 'over the wall' phenomenon of information processing (Cooper *et al.*, 1998), which leads to information related to a particular stage of the design or construction process being handed over without interaction from one stage to the subsequent stage. This is compounded by a lack of role overlap and interaction between the professions leading to disjunction between stages and professions involved in the design and/or construction activity. The boundary between design and construction is the most typical boundary where information is passed 'over the wall', resulting in buildability problems. It can also occur within a stage where, for example, design changes are poorly communicated to other professionals leading to wasted work on obsolete design versions. The problem can also arise in the production of finalised as-built drawings.

The related issue of design change was repeatedly highlighted as one of the biggest sources of delay in construction projects (Assaf *et al.*, 1995). Concepts of design freeze were discussed by the industrial team members, the eventual preference being for a progressive fixity of elements of design to allow concurrent advancement of the design and pre-assembly planning to be maximised. This is supported by the call for a 'total design paradigm' (Jamieson, 1997) which could be addressed by the co-ordinated extension of the design role involvement into the pre-project and post-completion phases of the product life cycle (Cooper, 1988). Poor buildability was also identified as a consequence of uncontrolled design changes.

Process problem factors identified for the construction phase

A commonly cited distinction of construction is site-based production (Koskela, 1992). Technical innovations originating in prefabricated and standardised products continue to be under-realised due to the need for a transparent and consistent process which will allow the repetition of benefits to the overall process to be achieved.

Table 3.1 addresses problem factors identified at each of the stages of the design and construction project and identifies solution mechanisms used in developing the Process Protocol at both a philosophical and a process-mapping level.

The concept of the Process Protocol: the drivers and philosophies

Having understood the problems as represented by the literature and industry it was recognised that the concept of the Process Protocol should consider the following points:

- The need for a model which 'is capable of representing the diverse interests of all the parties involved in the process . . . to be able to provide a complete overview'.
- There will be no best way for all circumstances (Plossl, 1991) but a generic and adaptable set of principles will allow a consistent application of principles in a repeatable form.
- There was a need for a coherent and explicit set of process-related principles (Plossl, 1991) and a new process paradigm which could be managed and reviewed across the breadth and depth of the industry and

which would focus on changing and systematising the strategic management of the potentially common management processes in construction while accommodating the fragmentary production idiosyncrasies.

- There was a need for design and construction operations to form part of a common process best controlled by an integrated system.
- There was a need for a definable and repeatable Process Protocol which would allow an IT solution to be devised that would manage both the Protocol and its information, and would also allow systematic and consistent interfaces between the existing practices and IT practice-support tools to be operated. Simplicity in the protocol and its operation are essential (Plossl, 1991). There should be clarity in terms of what is required, from whom, when and with whose co-operation, for whom, for what purposes and how it will be evaluated.
- There was a need for standardisation of the deliverables and the roles associated with achieving, managing and reviewing the process.
- There was a need for an industry-wide co-ordinated process improvement programme.
- A clear plan for future IT was needed to support the development of a repeatable and generic protocol.
- A philosophy of early entry into the process for the key functionaries must be assured with an emphasise on design and planning to minimise error and reworking during construction. It should be an extended process with earlier entry to allow a co-ordinated and recognisable/manageable professional contribution to the requirements capture and pre-project phases of client project planning – termed the pre-project phases.
- There should be an extension of the recognised construction industry involvement in the process beyond completion – a post-completion phase.

The philosophies: The process was therefore developed with the following philosophies in mind:

- Define and manage the process using a set of modular process phases which can operate sequentially and/or concurrently (and in small-scale projects may be combined so long as the key functional and function-driven deliverables and activities are maintained). The phase modules to comprise:

Pre-project phases
Demonstrating the need (phase 0)
Conception of need (phase 1)
Outline feasibility (phase 2)
Substantive feasibility study and outline financial authority (phase 3)

Table 3.1 Factors affecting the construction industry and process solution mechanisms.

Pre-project phases

Problem factors	Solution mechanisms
'The starting point for improving construction is to change the way of thinking, rather than seeking isolated solutions to the various problems at hand' (Koskela, 1992).	Change the culture of construction across the entire depth of the management hierarchy.
'There is no one best way to control manufacturing' (Plossi, 1991).	There will be no one best way to control all design and construction circumstances, but a generic and adaptable set of principles will allow a consistent application of principles in a repeatable form.
The lack of a convention of being involved in pre-project consultation.	Earlier and fuller involvement of design and construction professions in pre-project advisory role, possibly facilitated through an implicit or explicit client/ designer or client/contractor strategic partnering arrangement.
Need to foster better relationships (Jamieson, 1997).	Strategic partnering to provide opportunities to enter the client process of project planning and requirements capture. In subsequent phases of the design and construction process, the use of team building and virtual company techniques to establish and maintain team relationships at and between the managerial and operational levels throughout and/or beyond individual projects.
Contract-led restrictions on early involvement.	Alteration of contractual arrangements retention and/or consultative fee basis, possibly facilitated through an implicit or explicit strategic partnering arrangement.
Assessment of risk not being done as an overt or process-based exercise; tends to be done by individuals (Shahat & Rosowsky, 1995).	Risk assessment as explicit pre-project task. Establish a clear protocol for the early addressing of risk and risk allocation by the stakeholders in the pre-project process phases.

Pre-construction phases (1)

Problem factors	Solution mechanisms
Compatibility of processes (Hinks *et al.*, 1997).	A generic process that includes management processes and IT support, integration processes and process application protocols and specifications.
'Manufacturing operations forming parts of a common process are controlled best by an integrated system' (Plossi, 1991).	Design and construction operations forming parts of a common process are controlled best by an integrated system.
Linearity and sequential nature of existing design and construction examples in manufacturing and NPD 'process' (Jamieson, 1997).	Concurrent protocol required (Cooper, 1994).
Inconsistency of existing processes (Jamieson, 1997). Lack of concurrent managerial applications (Jamieson, 1997).	Adopt principles of concurrent engineering within a high-level process protocol.
Subsequent difficulties in producing and using construction production information (Mokhtar & Beddard, 1995).	Use of a product data modelling approach to production planning and information management (Mokhtar & Beddard, 1995).
	Co-ordinated data exchange protocols and a *change management* role to control changes in product or process – an explicit information management protocol linked to a consistent process protocol could be supported by an IS tool.
	Clear deliverables for functions and activities managed in a consistent and comprehensive information protocol linked to a supporting process management protocol.
	The function-driven deliverables for each process phase to be transparent and simply defined.
Poor client briefing (Jamieson, 1997). Accessing the client brief (Jamieson, 1997). Compounded by an excessive but unco-ordinated and unstructured bureaucracy in client bodies (Assaf *et al.*, 1995).	Earlier and fuller involvement of client in briefing phase. Address the client briefing process and roles.

Continues

Table 3.1   *Continued.*

Pre-construction phases (1)

Problem factors	Solution mechanisms
Poor co-ordination in design or project planning phases (Jamieson, 1997).	Process and IT support integration protocols.
Design change is one of the biggest sources of delay in construction projects (Assaf *et al.*, 1995).	Establish 'a generic management mechanism' (Hannus *et al.*, 1995).
Poor buildability consequences.	Collect and apply information to support the management process and co-ordination activity, to include the process decisions and operation as well as the production and design data. Especially if integrated at all levels (Hannus *et al.*, 1995).
	Co-ordination of functions and activities (Jamieson, 1997).
	Make the process directly observable and plottable (Koskela, 1992).

Pre-construction phases (2)

Problem factors	Solution mechanisms
Operate a total design paradigm (Jamieson, 1997).	Establish product design specifications (Jamieson, 1997) using generic activity, deliverable and function statements for strategic process management application.
	Operate the principle of concepts of progressive fixity of elements of design to allow concurrent advancement of the design and pre-assembly planning certainty to be maximised.
	Support with a single point information management and change management authority.
	Support with the extension of the design role involvement in the pre-project and post-completion phases of the product life cycle (Jamieson, 1997; Cooper, 1988).
	Emphasise effort on design and planning to minimise error and reworking during production.

Pre-construction phases (2)

Problem factors	Solution mechanisms
Unclear or missing information (BRE, 1995). Complex information flows (Jamieson, 1997).	The creation of pre-planned design information requirements which can be used to co-ordinate the brief-capture activity; also in creating design outlines and specifications.
	Recognise the iterative nature of information creation and use (Jamieson, 1997).
	This to be supported by a storage, retrieval and application protocol and mechanism for information gathered from previous project or created for the specific project to be reviewed and passed on to future projects – a legacy archive (Hinks *et al.*, 1997).
IT, process management, and communication are inter-dependent (Hinks *et al.*, 1997). Improved communication required. 'Over the wall' information handling.	Improved communication protocols and a facilitating role to allow communication to be guided. *Meta communication* planning needed (communication about communication expectations).
Information is not being widely used for strategic application purposes (Jamieson, 1997).	Strategic use of process information support with an organised and explicit information management role.

Pre-construction phases (3)

Problem factors	Solution mechanisms
Assessment of risk not being done as an overt or process-based exercise; tends to be done by individuals (Shahat & Rosowsky, 1995).	Risk assessment as explicit pre-project task.
	Support with the establishment of a phase review board comprised of stakeholders with early process entry to represent the virtual company, and assess phase progress and adjudicate on process continuation.
Lack of follow-on of risk assessment and decision data between phases (Jamieson, 1997). Lack of review and feedback of risk decision data analysis.	Establish overt risk assessment activity concurrency and overlap of risk assessment and assessments.

Continues

Table 3.1   *Continued.*

Pre-construction phases (3)

Problem factors	Solution mechanisms
Consequent compounding of errors throughout process (Jamieson, 1997).	Storage and review of decision data.
	Scope for construction professionals to assist clients undertaking pre-project feasibility analysis.
	Establish a clear protocol for the early addressing of risk and risk allocation by the stakeholders in the pre-construction process phase, including handover of risk assessment data between phases.
Professional conventions/boundaries. A need for a clear process management role and authority vested in a single individual (Male, 1996).	Professional roles and inter-relationships within the design and construction process need to be defined and/or revised. Responsibilities for Functional Activities in projects to be defined, and integration through managed team approaches.
	Establish a process manager or broker role (Male, 1996).
	Minimise the amount of control information required during the achievement of process deliverables by having a clear and transparent set of definitions of deliverables and allow professional autonomy in achievement of deliverables. Co-ordinate using a process management role operating within discrete modular phases and acting as a management link between phases.

Pre-construction phases (4)

Problem factors	Solution mechanisms
Need to exchange information across professional boundaries (Hannus *et al.*, 1995). Consequent under exploitation of potential cross-fertilisation of ideas (Jamieson, 1997).	Use virtual teams for multiple projects.
	Establish protocols for data ownership, management and use.
	Establish a project information management role and protocol.
	Establish function-based teams to achieve the deliverables required for each distinct phase of the process. Minimise the interdependency of these teams, maximise the communication channels and stipulate the information that must be communicated, how and to whom (parties and functionaries). Operate via a change management role for information management, which also acts as a communication link between phases (and to a legacy archive).

Pre-construction phases (4)

Problem factors	Solution mechanisms
Non-repetitive, complex and dynamic characteristics of the construction industry (O'Brien, 1997). Need to achieve control of the entire process (Koskela, 1992).	Establish a generic protocol to allow process management to be repeated without constraining the complexity and dynamic nature of production. Control the process so that the changes are in product not process, thereby retaining consistency of management control. Use controlled process to enhance capability to accommodate product changes.
The need for generic process models (Jamieson, 1997; Whitelaw, 1996; Aouad *et al.*, 1994; PISTEP, 1994).	It is essential that the parties involved in project teams are operating to a consistent process which requires broadly similar process capabilities of the various team members. This will mean that industry-wide process improvement is required from the design and construction process to achieve repeatability and hence manageability (Hinks *et al.*, 1997). This means linking the professional capabilities in the pre-construction phases as well as the production phase parties, and linking capability and process involvement across the phases. Embody the process and its constituent roles and activities in clear protocol guides.
Incompatibility of the construction IT.	Generic document structures and the development of an integrated document data environment (Armstrong *et al.*, 1995). Neutral interchange and integration of documents is needed (Armstrong *et al.*, 1995). Data to be systematically gathered from previous projects (and improved following feedback) using an appropriate feedback mechanism – a legacy archive (Hinks *et al.*, 1997), application of IT for the facilities management of the building using data on this project beyond completion (Kohler *et al.*, 1995). Application of STEP (Jamieson, 1997; Armstrong *et al.*, 1995) and object modelling approaches (Aouad *et al.*, 1994; Armstrong *et al.*, 1995; Aouad *et al.*, 1993). Provision of a number of technical and product-orientated support provisions which are required.

Continues

Table 3.1   *Continued.*

Pre-construction phases (4)

Problem factors	Solution mechanisms
Inconsistency in IT representation of components and objects.	Representation of processes and objects requires to be consistent (Armstrong *et al.*, 1995; Hannus *et al.*, 1995).
	A protocol for the process integration of IT and data.
	Application of STEP (Jamieson, 1997; Armstrong *et al.*, 1995) and object modelling approaches (Aouad *et al.*, 1994; Armstrong *et al.*, 1995; Aouad *et al.*, 1993).
Data availability (O-Brien, 1997) in the early stages of the process where information is unstructured.	A Protocol for the process integration of IT and data (Lockemann *et al.*, 1995). This to be supported by a storage, retrieval and application protocol and mechanism for information gathered from previous projects or created for the specific project to be reviewed and passed on to future projects (or this project beyond completion) (Kohler & Bedell, 1995) – a legacy archive (Hinks *et al.*, 1997).
Application of STEP (Jamieson, 1997; Armstrong *et al.*, 1995) and object modelling approaches (Aouad *et al.*, 1994; Armstrong *et al.*, 1995; Aouad *et al.*, 1993).	Support with loading of planning effort to the pre-construction and pre-project phases of the process to allow ambiguity and late stage design changes to be minimised, thereby allowing greater pre-production planning and a smoother, more controllable production phase. Enhanced predictability of production would allow the opportunity of prefabrication and standardisation in the product to be realised.

Pre-construction phases (5)

Problem factors	Solution mechanisms
Poor project planning (Jamieson, 1997).	Adopt manufacturing new product development methods and techniques (Jamieson, 1997).
Poor process control and definition.	Establish process control protocols and single point authorities for management of process.
	Develop protocols for the evaluation of process performance and improvement Jamieson, 1997).
	Establish phase review board system (Cooper, 1993).

Construction phases (1)

Problem factors	Solution mechanisms
Productivity lags behind that of manufacturing (Koskela, 1992).	Focus on errors at source (including design) using a feedback and review mechanism in a structured process involving professional integration. Focus on JIT philosophy of process improvement and checking at source rather than quality programme approach of TQC school (Koskela, 1992).
Scope for technical innovations originating in prefabricated and standardised products continues to be under-realised.	Requirement for a consistent process to accommodate and stimulate the uptake of product and production innovations. Focus on people as well as technological innovation. People are the process drivers.

Pre-construction phases
Outline conceptual design (phase 4)
Full conceptual design (phase 5)
Co-ordinated design, procurement and full financial authority (phase 6)

Construction phases
Production information (phase 7)
Construction (phase 8)

Post-completion phases
Operation and maintenance (phase 9)

- Create a structured protocol for gathering, collating, storing, managing, retrieving and applying feedback in the product and process to allow an individual project to benefit from a recording of its own process; and a longer-term scope to produce learning cycles to improve products, production processes and management processes.
- Focus on the perspective of process management rather than project management.
- Minimise the process management intervention harmonic by forcing functional teams to co-operate.
- Recognise *people* as process drivers rather than simply the technological innovation that drives production.
- Make education a continuous effort (Plossl, 1991).
- Design the implementation of the new protocol to be incremental and to embody a continuous improvement paradigm within the Process Protocol itself (review and feedback).

- Control the process so that the changes are restricted to the product and not the process. This would allow consistency of management control to be retained. Use a consistent and controlled process which would enhance capability to accommodate product changes.
- Benchmark the process in operation using a set of broad evaluation criteria (Koskela, 1992). Focus on causes of process and product quality rather than simply results such as costs.
- Load planning effort to the pre-construction and pre-project phases of the process to allow ambiguity and late-stage design changes to be minimised thereby allowing greater pre-production planning and a smoother, more controllable production phase.
- Enhance predictability of production to allow the opportunity for pre-fabrication and standardisation in the product to be realised.
- Incorporate a clear protocol for the early addressing of risk and risk allocation by the stakeholders in the process.
- Ensure an extended involvement of all stakeholders early in the process and throughout the process, especially those exposed to risk.
- Ensure the establishment of a virtual company for the process.
- Establish a phase review board comprised of stakeholders with early process entry to represent the virtual company and to assess phase progress and adjudicate on process continuation.
- Create a model which is non-specific to a particular industry sector's perspective (Jamieson, 1997).
- Develop a model which is generic enough to allow local customisation to the project, professional and operational circumstance without losing generality – there should be a high-level set of strategic guidelines.
- Devise a protocol to allow strategic management control without constraining flexibility of product or production innovation.

Operation of the Protocol

As a result of considering the conceptual aspects of the process the following aspects were built into the development of the Process Protocol:

- The adoption of a single point of contact who has the authority to orchestrate the process.
- That the single-point authority should be defined and chosen according to capability, not profession. Therefore the process uses multi-skilled professional virtual teams operating as function-driven process teams:
    Development management
    Project management

Resource management
Design management
Production management
Facilities management
Health, safety, statutory, and legal management
Process management
Change management

- Scope for supple decision-point system (gates) to be consistently hard (requiring a firm decision for project progression and a recognised and consistent decision point in the industry process) or soft (which would allow the focusing of team efforts and facilitate concurrent process planning and design across decision points which by their nature tend to vary in certainty and prescription).
- Make the process directly observable and plottable (Koskela, 1992).
- Embody the process and its constituent roles and activities in clear protocol guides.
- A defined set of function-driven deliverables for each process phase to be transparent and simply defined.
- A defined set of deliverable-driven activities. Minimise the amount of control information required during the achievement of deliverables by having a clear and transparent set of definitions of deliverables and allow professional autonomy in the achievement of deliverables. Co-ordinate using a process management role operating within discrete modular phases and acting as a management link between phases.
- Develop integration protocols for existing IT and integrated IT and IS to support and enable process management and process improvement. Couple production-support IT with process-support IT. Develop IT and IS which link the professional roles and facilitate capture and re-use of knowledge in the early stages of the process.
- The concept of a legacy archive in which process phase and process management information is systematically stored and retrieved as feedback and process management support information.
- Re-application of project and process information to future projects using a legacy application tool.

The Process Protocol described

The last section described the conceptualisation of the Process Protocol based on contemporary problem areas in the construction sector. It has also set the scene for developing a framework that provides potential solutions to the issues raised. Here we describe the essential elements of the generic design and construction protocol.

Key principles

As a result of the initial review of the literature and the identification of the industry's requirements, the drivers and philosophies, six key principles were considered to provide the basis for an improved process.They are drawn heavily from the manufacturing sector where process thinking and continuous improvement have been focused upon since the 1970s. In addition, many of the principles relate to recognised problem areas in construction where significant improvements have been called for (*inter alia* Banwell, 1964; Latham, 1994). The six principles are as follows.

(1) Whole project view

In the construction industry the definition of a project has traditionally been synonymous with actual construction works. As such the pre-construction and post-construction activities have been sidelined and often accelerated to reach the next construction stage or to move on to the next project. The result of this is an insufficient identification of the client's requirements and a delayed exposure of the potential solution to the internal and external specialists. Any contemporary attempt to define or create a 'design and construction process' will have to cover the whole 'life' of a project from the recognition of a need to the operation of the finished facility. This approach will ensure that all issues are considered from both a business and a technical point of view. Furthermore this approach recognises and emphasises the inter-dependency of activities throughout the duration of a project and also focuses on the 'front-end' activities so that attention is paid to the definition and evaluation of client requirements in order to identify suitable solutions.

(2) A consistent process

During the review of existing models and descriptions of the design and construction process it was quickly established that little consistency existed. In such an environment, the problems encountered by temporary multi-organisations (TMO) can be compounded. Luck & Newcombe (1996) support this view, describing the 'role ambiguity' commonly associated with construction projects.

   Development of this generic Process Protocol provides the potential to establish its consistent application. Through consistency of use the scope for ambiguity should reduce. This, together with the adoption of a standard approach to performance measurement, evaluation and control, should facilitate a process of continual improvement in design and construction.

(3) Progressive design fixity

The 'stage-gate' approach found in manufacturing processes (Cooper, 1994; see Chapters 1 and 3) applies a consistent planning and review procedure throughout the process.

Phase reviews are conducted at the end of each phase with the aim of reviewing the work executed in that phase, and approving progress to and planning the resourcing and execution of the next phase. Cooper, in his third generation process, saw the need for 'conditional-go' decisions at phase gates to accommodate aspects of concurrency. This philosophy is translated in the development of the protocol's phase gates. Phase gates are classed as either 'soft' or 'hard', with the soft gates allowing the potential for concurrency in the process while ensuring that the key decision points in the process are respected.

The potential benefit of this approach is fundamentally the progressive fixing and/or approval of information throughout the process. As Cooper (1994) states, the discipline of the phase review activity improved the conventional chaotic, *ad hoc* approach of manufacturing to which the construction industry of today could be compared.

(4) Co-ordination

Co-ordination is one area in which construction is traditionally perceived to perform poorly. This perception is supported by Banwell (1964) and Latham (1994) in addition to many other reviews of the industry. The need for improved co-ordination was also highlighted by the interviews with senior managers undertaken during the research project.

It is therefore proposed that co-ordination of the Process Protocol is undertaken principally, by the process and change management *activity zones*. Appointed by the client, the process manager will be delegated authority to plan and co-ordinate the participants and activities of each phase throughout the process. The actions of the process manager are supported by the change manager through whom all information related to the project is passed. In this role the change manager acts as the official interface both between the *activity zones* in the process and ultimately with the legacy archive.

(5) Stakeholder involvement and teamwork

It has been recognised in the manufacturing industries that the establishment of multi-functional teams in a development process reduces the likelihood of costly changes and production difficulties later on by enabling design and manufacturing decisions to be made earlier in the process.

(6) Feedback

The phase review process facilitates the recording, updating and use of project experiences. The creation, maintenance and use of a legacy archive will aid a process of continued improvement in design and construction.

The Process Protocol elements

The Process Protocol model is presented in Figs 3.1 to 3.4. Essentially the model breaks down the design and construction process into ten distinct phases. These ten phases are grouped into four broad stages, namely pre-project, pre-construction, construction and post-construction. These are described below.

Pre-project stage

The pre-project phases relate to the strategic business considerations of any potential project which aims to address a client's need. Throughout the pre-project phases the client's need is progressively defined and assessed with the aim of:

(1) Determining the need for a construction project solution.
(2) Securing outline financial authority to proceed to the pre-construction phases.

In currently acknowledged models of the design and construction process (*inter alia* RIBA, 1980; British Property Federation, 1983; Hughes, 1991 provides a comprehensive review) and recently published client-focused guides (CIRIA, 1995) this stage of a project is given scant consideration compared to the latter stages. However, these models assume that when approaching the construction industry, clients have already established 'the need'. While there is little evidence to suggest this is not the case, it would

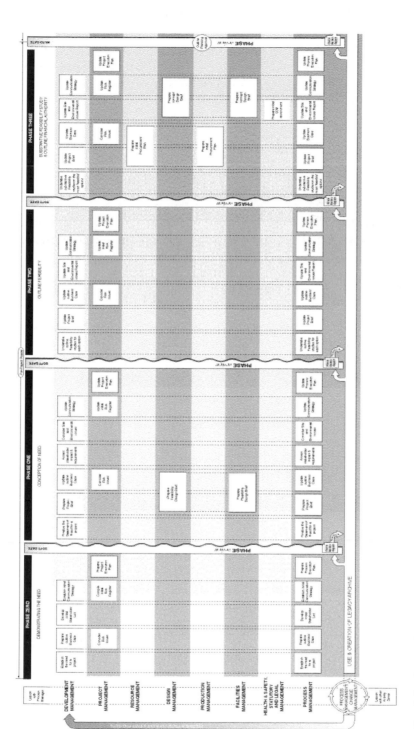

Fig. 3.1   Pre-project section of the Process Protocol.

seem reasonable to assume that the knowledge possessed by speculative building developers and consultants could assist any client in these early stages of a project. The problems associated with the translation of this need through the conventional briefing stage of design (O'Reilly, 1987) could be substantially eliminated via such an approach.

Pre-construction stage

With outline financial approval obtained the process progresses through to the pre-construction phases where the defined client's need is developed into an appropriate design solution. Like many conventional models of the design process the pre-construction phases develop the design through a logical sequence with the aim of delivering approved production information. The phase review process, however, adds the potential for the progressive fixing of the design, together with its concurrent development, within a formal co-ordinated framework. Progressive fixity should not be confused with 'design freeze' although to some this may be a desired aspect of the process. The major benefit of the fixity of design is the potential for improved communication and co-ordination between the project's participants as they pass through each phase. Given the dynamic market conditions which influence many construction clients' decisions, the need for flexibility must be addressed by the industry.

At the end of the pre-construction phases the aim is to secure full financial authority to proceed. Only upon such authority will the construction phase commence and this decision will be easier to make where the extent of the works and the associated risks can be readily understood.

Construction stage

The construction phase is solely concerned with the production of the project solution. It is here that the full benefits of the co-ordination and communication earlier in the process may be fully realised. Potentially, any changes in the client's requirements will be minimal as the increased cost of change as the design progresses should be fully understood by the time on-site construction work begins.

The 'hard gate' that divides the pre-construction and construction phases should not prevent a 'work-package' approach to construction and the associated delivery time benefits this brings. As with all activities in the process, where concurrency is possible it can be accommodated. The hard and soft gates that signify phase reviews merely require that approval is granted before such an activity is carried out.

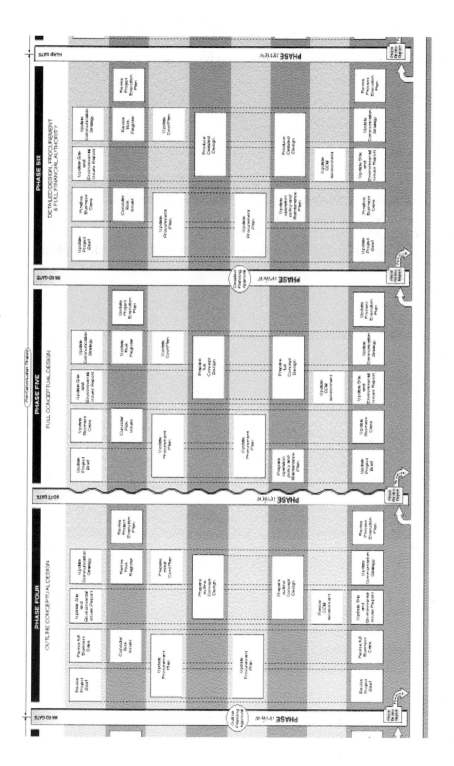

Fig. 3.2 Pre-construction section of the Process Protocol.

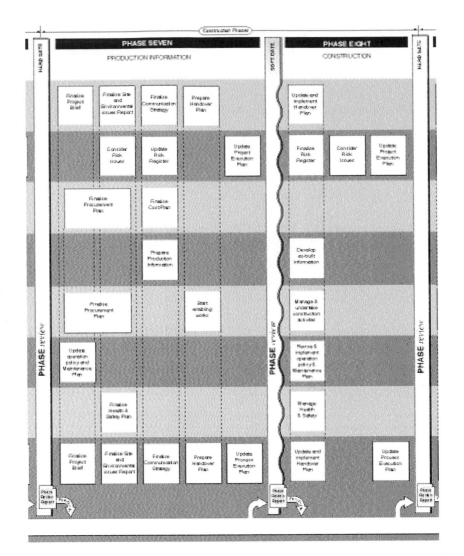

Fig. 3.3   Construction section of the Process Protocol.

Post-construction stage

Upon completion of the construction phase, the Process Protocol continues into the post-construction phases which aim to continually monitor and manage the maintenance needs of the constructed facility. Again, the full involvement of facilities management specialists at the earlier stages of the process should make the enactment of such activities less problematic. The need for surveys of the completed property, for example, should be avoided as all records of the development of the facility should have been recorded by the project's legacy archive.

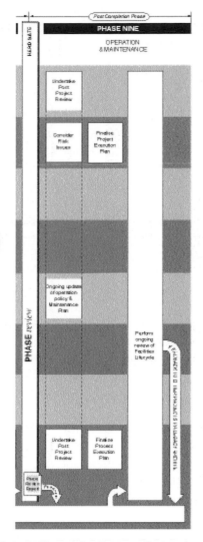

Fig. 3.4   Post-construction section of the Process Protocol.

The Activity zones

The earlier involvement of the project's participants throughout the process is a significant development of the conventional approach to building. Traditionally, a construction project's participants are referred to by their professional or expert status. Ball (1988) demonstrates how this may be attributed to the inherent class relations associated with each of the professions and expert groups. As with all class distinctions, the effect of this basis for organisational structure in design and construction is division.

A consequence of this traditional approach, in which even the more recent forms of contract procurement (design and build, management contracting, etc.) are included, is the poor communication and co-ordination commonly associated with construction projects.

The participants in the Process Protocol are referred to in terms of their primary responsibilities and are represented on the *y*-axis of the process model. It is recognised that traditionally, project to project, organisational roles and responsibilities change resulting in ambiguity and confusion (Luck & Newcombe, 1996). By basing the enactment of the process upon the primary responsibility required, the scope for confusion is potentially reduced and the potential for effective communication and co-ordination increased. The Process Protocol groups the participants in any project into 'activity zones'. These zones are not functional but multi-functional and they represent structured sets of tasks and processes which guide and support work towards a common objective (for example, to create an appropriate design solution).

A single person or firm can carry out an activity zone in small projects but in large and complex projects an activity zone may consist of a complex network of people and relevant functions and/or organisations. Since they are multi-functional, membership of the 'zones' is determined by the specific project task and/or process. For example, design management often has important input in the production management and facilities management activity zones amongst others, and *vice versa*.

Of the activity zones associated with the model, not all will be discussed here in detail. (Further details can be found in the Appendix.) Most of the zones are self-explanatory. However, the role of the process/change management and development management activity zones will be described as they present a significant departure from the conventional view of the design and construction process.

Development management

The development management activity zone is fundamentally the client/customer for the potential project. In the Protocol scenario, it is ultimately responsible for the success or failure of the project. Representing the major stakeholder in the process, it has an important role. It is via the brief prepared by the development management that the client/customer's needs are presented and ultimately interpreted. The development management is the only constant 'player' in the process. All other activity zones potentially consist of a dynamic membership as the needs of the project develop throughout the process. The extent to which the other participants in the process, particularly the process management, have authority to proceed is delegated by the development management. It is they who will ultimately review the work of the project's participants and sanction progress or cessation.

Development management is responsible for creating and maintaining a business focus throughout the project which satisfies both relevant organisational and stakeholder objectives and constraints. For example, a proposed speculative office development needs to satisfy the developer's objectives (say, return on capital) and constraints (say, available finance) as well as fulfilling other stakeholder considerations (say, compliance with prevailing planning concerns).

Project management

The responsibility for the effective implementation of the project to agreed performance measures rests with the project management activity zone. It is an agent of the development management activity zone for achieving their business and project requirements as set out in the business case and project brief.

The project management role is fairly well defined within the construction industry as traditionally those persons or parties who are responsible for the implementation not only of the project but of the 'process' as well. Although this method has proved to be effective in some circumstances it lacks predictability as the 'process' depends on the methods and beliefs of the project manager. The Process Protocol places the project management role within the framework of the design and construction process.

Typical responsibilities of the project management activity zone will include identification of project activities and deliverables; formulation of effective project execution plans; co-ordination of the project team towards satisfying the requirements of the client; having an input to the majority of the project deliverables; and liaising with process management throughout the process.

Resources management

Resources management is responsible for the planning, co-ordination, procurement and monitoring of all the financial, human and material resources of the project. Although the overall project budget is decided by development management, resources management is responsible for ensuring that all cost estimates and purchasing of goods and services are consistent with the requirements set by project management.

Design management

Design management is responsible for the design process which translates the business case and project brief into an appropriate product definition. It guides and integrates all design input from other activity zones.

Production management

This activity zone is responsible for ensuring the optimal solution for the buildability of the design, the construction logistics and organisation for delivery of the product.

Facilities management

Facilities management is responsible for ensuring the cost-efficient management of assets and the creation of an environment that strongly supports the primary objectives of the building owner and/or user.

Health and safety, statutory and legal management

Health and safety, statutory and legal management is responsible for the identification, consideration and management of all health and safety, statutory and legal management aspects of the project.

Process/change management

The process and change management activity zones are essentially the interface between the development management and the other project participants. Process management has a role independent of all other activity zones. A distinction must be made between this conventional view of a project manager and the process management role. Process management, as the title suggests, is concerned with the enactment of the *process*, rather than the *project*. Key to the success of each phase in the process is the production of project deliverables (reports and documentation associated with each phase). In this respect the process management is responsible for facilitating and co-ordinating the participants required to produce the necessary deliverables. Acting as the development management's 'agent' it will ensure the enactment of each phase as planned, culminating in the presentation of the deliverables at each end of phase review.

The change management function is further distinct from the process management zone as this role solely concerns (as its name also suggests) the management of change(s) which occur during the process. As the project becomes increasingly defined as each phase is enacted, changes (or rather updates) to the information required for the development of the project will be produced. These updates will be contained within the work required to develop the deliverable documentation associated with each phase. With respect to this, the change management (CM) activity zone facilitates the holding, review and dissemination of all this information as the project progresses.

It is within the change management function that IT potentially plays a fundamental role. Given the vast amount of information generated throughout a project's life cycle (Aouad *et al.*, 1994) and the need for its quick and effective dissemination, IT may offer a suitable solution. However, the need for judgement and discretion, especially in the earlier strategic phases of the process, will always involve the development management's intervention and this alone is likely to prohibit the use of IT as a total solution.

Deliverables

Each gate of the Process Protocol represents a decision-making point and the decisions are based primarily on deliverables – documented project and process information. They are compiled by project management to form the phase review report which includes all the deliverables specific to the phase and as they are defined by the Process Protocol for the specific project. This phase review report forms the basis for the client body (i.e. development management) to make a decision concerning the future of the project.

The deliverables are 'live' documents which change throughout the majority of the process. They can be in one of the following states:

(1) *Initial*: preliminary information is presented.
(2) *Updated*: current information is updated.
(3) *Revised*: major changes/decisions will significantly alter the content and context of the deliverable.
(4) *Finalised*: the information presented is agreed and it is unlikely to change throughout the duration of the project.

The Process Protocol includes a number of likely deliverables which are briefly outlined below.

- *Stakeholder list*: Stakeholders are those persons and/or organisations whose views, interests and/or requirements can have an impact on or are impacted by the initiation and/or formulation and eventual implementation of the project solution. When a large number of stakeholders are considered they might be prioritised to illustrate their importance and involvement from a client's perspective in the proposed project.

- *Statement of need*: The client's needs should be clearly identified and defined as the project progresses to include a greater amount of detail. This deliverable should aim to provide the project team with a succinct indication of the clients' reasons for the potential project.

- *Business case*: The business case should consider the risks, costs and benefits associated with any proposed solution from a number of perspectives related to finances, product/service and customers.

- *Project execution plan*: This deliverable will have to illustrate a clear understanding of the requirements of the project execution. This will potentially improve the chances for the successful implementation of the project. Together with the rest of the deliverables, the project execution plan (PEP) is a 'live' document which changes in content primarily as a result of design changes for the project solution.

- *Process execution plan*: In relation to the PEP, the process execution plan focuses on the project process requirements that will potentially increase visibility and enable the production of the respective phase deliverables for submission at the phase review meeting.

- *Performance management report*: The effective implementation and monitoring of any project will have to depend on effective performance measurements which form the performance management report. The measures applied to the project will be of a diverse nature including financial and technical as well as human resources, materials handling, IT use and utilisation, etc.

- *Communications strategy*: Effective communications can potentially reduce lead times, improve quality and ensure accurate and prompt informing of everybody involved in the project. The communications strategy report should clearly indicate the means of communicating for the project team. This might include the use of AutoCAD, email, intranet, etc.

- *Procurement plan*: The success or failure of a project could depend on providing the right resources at the right time and at the right location. Indeed, the procurement of services, products, finance and programme, almost irrespective of the method used, could have a great influence on the project's outcomes.

- *CDM assessment*: The Construction (Design and Management) Regulations (1994) place new duties on clients, planning supervisors, designers and contractors to plan, co-ordinate and manage health and safety through all stages of a construction project. Since it is the client's responsibility to comply with the CDM regulations, provisions for reporting on those issues should be made.

- *Project brief*: This document should mainly identify the scope of the project and more specifically that of the proposed solution(s). For the majority of the process the project brief remains a 'live' document.
- *Design brief*: The design brief should document the different project solutions prior to the enactment of the feasibility studies. The information presented should be used to form part of the initial project brief and the updated business case.
- *Concept design plan*: This should include the results of the feasibility studies with an identification of the proposed project solutions, likely timescales and resources necessary to carry out the work.
- *Outline concept design*: This deliverable will include the final site for the construction of the project solution(s). It aims to inform the business case with regard to the form, function, specialist requirements and programme likely to be associated with the proposed solution(s).
- *Full concept design*: It should identify the major design elements of the proposed solution and should be architecturally detailed so that a submission for detailed planning approval can be made.
- *Product model*: The product model should include all the major design elements of the single solution so that the detailed design work can be carried out. It can be presented as *co-ordinated* – structural, mechanical and electrical components should have a high level of technical detail with corresponding specifications for the major design elements; and *operational* – the design is represented in terms of work packages so that construction works can begin.
- *Cost plan*: The cost plan presents the potential and actual costs for the construction project such as cost/benefit analyses, cash flow requirements and value engineering.
- *Maintenance plan*: The maintenance needs of the finished facility/project should be considered.
- *Production process map*: In cases where the production of the project solution is phased the production process map should indicate the phasing strategy such as timing and resources.
- *Hand-over plan*: When the facility is handed over to the operation and maintenance teams the hand-over plan should include as-built drawings, services and operations information, commissioning information, etc.

In addition, a deliverable which is inherent to the whole of the Process Protocol is the risk study/assessment which should be undertaken at every phase and its outcomes considered at every gate to ensure the seamless continuation of the project. Those risks identified can be further prioritised and contingency plans put in place to overcome or minimise them.

Legacy archive

The Process Protocol aims to utilise IT and communication technologies to ensure the timely storage, transfer and retrieval of project information throughout the duration of the project's life cycle. The legacy archive is a 'live' communications link which does not just facilitate co-ordination on a project-based environment but offers the opportunity of learning from previous project successes and/or failures. Such tools are now available and have been discussed in Chapter 2.

IT and the Process Protocol

A fundamental part of the process is its accompanying IT map. The Latham Report (Latham, 1994), which was one of the main catalysts for the process, focused upon the fragmented nature of the industry and poor communication between the parties working on a construction project. It is acknowledged that in order to achieve the many potential benefits that will come with the improved Process Protocol and to eradicate the poor communication between parties involved, there needs to be a significant use of IT to support it (Aouad, 1997).

The IT map (see Fig. 3.5) is a 'vision for the future' of the use of IT in the construction industry and covers all the stages of the design and construction process identified in the Process Protocol, from establishing the need for a project to the operation and maintenance of a building.

Although there are many other 'vision for the future' pieces of research in this area the majority are mainly concerned with the latter stages of design and construction. The recent developments in IT technology in areas such as VR and 3D modelling which support rapid prototyping are an essential part of the process and hence are illustrated in the map.It is essential that the use of IT is set in the context of the whole design and construction process – both the front-end pre-project stages and the later production and post-completion stages. Indeed the industry must move more rapidly towards intergration of the technologies.

The use of technology during the process

At the pre-project stage of the process, phases zero and one, there is one main question being asked: 'Do we need a building?' To provide the answer to this question simulations should be utilised. This can be either a numeric or a visual format and should use tools such as VR, 3D modelling and economic analysis tools with the required data being obtained from an archive of previous projects.

During phase two – outline feasibility – a product-based case-retrieval system is recommended for use and this again could utilise data from an archive of previously undertaken projects in order to give a basic cost for the various project options.

Phase three, which is the substantive feasibility study, AI tools such as neural networks, knowledge-based systems and case-based reasoning could be used as an aid to enhance creativity in the initial production of the design while multi-media applications can help ease the distribution of information to laypersons such as the client. The second stage, the pre-construction stage, incorporates all the design stages, i.e. outline conceptual design, full conceptual design, co-ordinated design and also product information. At this stage a design needs to be established that fulfils the requirements identified in previous feasibility studies.

At the outline conceptual design phase the use of AI tools is continued from the previous feasibility phase as this area again can utilise the creativity enhancement properties of these tools.

During the full conceptual design and co-ordinated design phases the IT map recommends the use of cost-planning applications in order to ensure that design and construction costs do not exceed the budget.

Also during these stages VR, 3D modeling, 2D CAD and constructability/usability modeling tools are shown. The VR, 3D modeling and 2D CAD are used in the production of the design and design drawings. VR can also be used along with the constructability/usability modeling tools to see if the design is free from 'errors' by utilising features such as clash detection, and can help establish whether the building will be fit for its purpose once constructed.

A project planning application is shown at the co-ordinated design and product information stages, the purpose of which is the production of the project plan.

Stretching from construction stage phase seven (product information) to phase eight (construction) are four IT application areas. These are as follows:

- **3D modeling and VR** which is situated at the project board and business case functions. These applications are used here as visualisation tools for board members and higher management of the contractor and/or client so that they can interrogate the VR or 3D model for detailed information on individual elements of the building, see and monitor the progress of the project and interrogate the model for cost information.
- **Progress reporting** and as-built model generation, situated at the cost control function. This would most probably be used by the client and

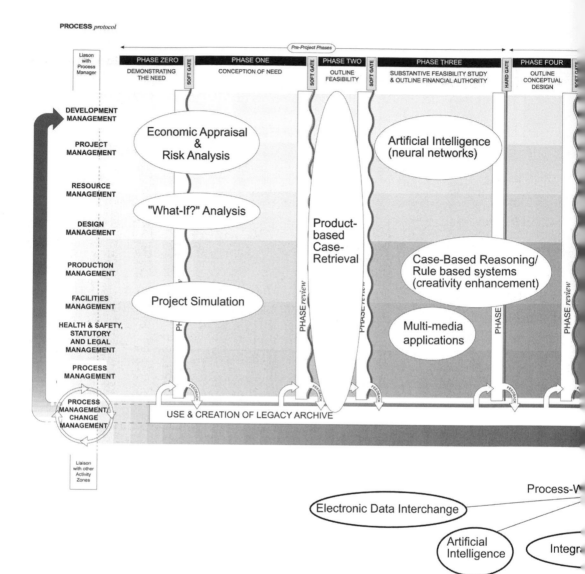

Fig. 3.5   The IT Process Protocol map.

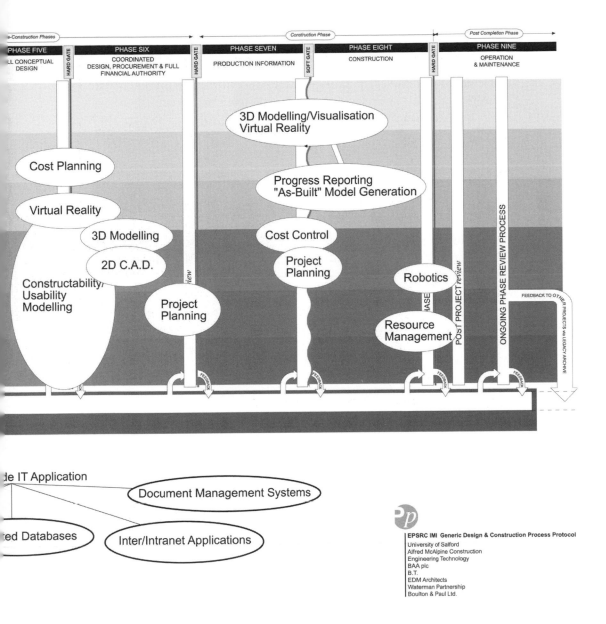

Fig. 3.5   *Continued.*

could also use VR or 3D modeling tools but would most probably utilise a generic accounting package or a project planning application that has good costing and progress monitoring facilities for the progress reporting and VR or 3D modeling for the as-built model generation.

- **Cost control** and **project planning**, situated at the design and construction function. This would most probably be used by the contractor and would utilise generic accounting packages and project planning applications to undertake cost control and planning of the project respectively. Two application areas at the construction stage phase eight (construction) and post construction stage phase nine (operation and maintenance) are shown on the IT map (see Fig. 3.5). At the design and construction and production functions robotics is shown which could be used for various construction and maintenance activities such as the positioning of pre-cast concrete flooring, the erection of cladding for building inspections and repair of various parts of the building, and cleaning.

For production and facilities management functions a resource management application area is shown. This would probably be a generic facilities management package that would assist in the running and maintenance of the building after construction.

In addition to the process-specific application areas the IT map identifies various IT application areas that should be used throughout the whole process protocol.

The five areas identified are as follows:

- Electronic data interchange (EDI).
- Inter/intranet applications.
- Document management systems.
- Artificial intelligence.
- Integrated databases.

These IT application areas are designed to be used throughout the Process Protocol in order to exchange information and support communication between all those involved in the project.

Electronic data interchange (EDI) and inter/intranet applications such as email, groupware and World Wide Web applications will help improve communications not only between the main parties involved in the project such as contractor, client and architect but also between, for example, suppliers and various legislative organisations. This improvement should therefore provide better co-ordination and management of the project. Document management systems could also help with communications as correspondence between the parties involved in the project such as faxes,

EDI invoices and email could be stored and retrieved at a later date for the settlement of claims.

The use of AI could help in various areas of the process. Technologies such as neural networks, case-based reasoning and knowledge-based systems could provide decision support systems that can manage and automate various processes within the protocol.

Concluding the process-wide IT applications, the use of integrated databases would allow the sharing and exchange of information between all parties involved in the project. An integrated database combined with the other technologies for communication and AI would provide a very powerful tool that could automate numerous processes such as project planning and bill of quantities generation.

As mentioned previously this is a vision for the future, i.e. a long-term objective, as various technologies mentioned are unlikely to become accepted and used by industry until further research and development has been undertaken on them.

Aouad (1997) suggests that when IT usage is similar to that shown on the IT map, it should be possible for the client to walk through and interrogate various aspects of the designed building, such as cost and specification using VR and also the information stored in the integrated database. It could be possible for the project manager to select the most appropriate procurement path using neural networks techniques. The designer could be able to select the most suitable design using case-based reasoning techniques from information stored within the legacy archive of the integrated intelligent database. With the use of the intelligent integrated database and improved communication applications the information flow between applications and project participants would be transparent and the data would remain unduplicated and uncorrupted.

Postscript: the development of sub-processes and a toolkit

Process Protocol level 2

Industry interest and acceptance of the framework provided the impetus for further research under the Process Protocol level 2. The first project concentrated on the high-level protocol and the second aimed to develop the sub-processes of the eight activity zones that exist within the original generic design and construction Process Protocol model. These sub-process maps provide an increased level of detail and description compared to the existing Process Protocol map. This section describes the methodology used to develop the sub-processes and explains the reasons for developing a bespoke modelling methodology rather than using the standard process modelling techniques. The methodology enables all of the information relating to the sub-processes to be represented as a series of process maps and when viewed holistically presents an integrated generic decomposition of the processes on the high-level map.

Sub-process map definition

A key requirement of the sub-process maps was that prior understanding of process modelling techniques should not be a prerequisite to understanding them. Process modelling tools such as the IDEF family were considered but were felt (as before) to be too complex for certain members of the targeted user group. The key was in the representation (Cheung, 1998) of the process and it was felt that none of the tools available met the project's requirements. Therefore it was necessary to develop an original process map template.

Visio Professional was used to design and create the sub-process maps. Visio had the ability to attribute information to objects within its diagrams and store this information within its own database. This led to the possibility of exporting data to other applications thus increasing the usefulness of the maps and perhaps assisting the process toolkit which was another primary deliverable of the project. The number of maps in a series is dependent on the level of detail attributed to a process and the number of processes in the activity zone.

A map was created that represented all of the information the project required. It was formally presented to the project partners and conditionally approved. Feedback about the map's format was received and internal development workshops were held to refine the map. Issues arose from the workshops regarding the modelling of the sub-process maps. There was a lack of information relating to the different levels of process modelling; for instance, the criteria used to distinguish between a level 2 and a level 3 process were not stated. It was decided that this distinction is a subtle one and relies on the experience and judgement of the process modeller.

The map produced was discussed and the consensus was that it needed to be simplified. The content of the maps was becoming complex and methods to reduce this complexity while still showing all the detail required were considered. A decision was made to only show level 1 deliverables.

Processes that were common to all or some of the activity zones were beginning to be identified and considered as generic process components. An issue that remained unsolved was how to illustrate the interaction of these intra-activity zone components. Process inputs and outputs were indicated at this stage by an arrow-headed line and labelled.

Information structure on the maps

A further workshop was held. Debate focused on the distribution of the content of the maps with respect to the validity of the content and whether it appeared at the correct phase and in the correct sequence. The detail of the level 2 and level 3 processes was considered as was the level in which the process should belong: should the process be moved up to level 2 or down to level 3 or excluded completely? This allowed the team to familiarise themselves with the content of the maps.

This workshop saw the introduction of phase grouping. Previously it had been noted that activities were often repeated through several phases. Therefore in order to simplify the content of the maps and to avoid repetition activities were grouped together logically. Whether these processes occurred at the beginning of a phase, during a phase or at the end of a phase would determine the standard group in which they would reside. A further workshop was held which was notable for two main outcomes. The first introduced the principle of process ownership. All of the processes would have an activity zone having overall ownership of a process to ease the co-ordination of a process. The second was the introduction of a new process symbol that illustrated which activity zones were participants in the process/sub-process. This solved the problem of how to show intra-activity zone processes. A single glance would now indicate the origin and ownership of a process and what activity zones participated in the process thus validating activity zone interfaces. It was agreed to incorporate the new symbol on to the developing process map template together with the activity groups.

An industrial workshop was held to gain industrial validation of the sub-process maps. This workshop considered the modelling rules and conventions used to define and model the high-level processes into sub-processes. A total decomposition modelling technique was adopted to illustrate the decomposition of a process and its sub-processes. Logical dependency is represented between processes when it exists.

Modelling rules

The aim of the modelling work is to provide a visual representation of the sub-processes of the activity zones of the Process Protocol map. This is achieved by illustrating 'what' are the sub-processes of the high-level processes identified in the Process Protocol map and 'how' these sub-processes interact. As a result it will be possible to provide models for individual phases.

Modelling conventions

The main convention types used for the modelling of the sub-processes of the Process Protocol map include:

- Phase start-up activities.
- Map title including phase number, phase title and activity zone name.
- Three process levels, to include a generic top level and two subsequent levels that represent the decomposition of the top-level processes.
- Ongoing activities.
- End of phase activities.
- Lexicon.
- Map title block including author, version, phase number and title, etc.

The phase start-up, ongoing and end of phase activities are illustrated on separate process maps. They represent activities common to many activity zones and represent activities undertaken for most phases.

Process representation

The processes and sub-processes are denoted by using the symbol shown in Fig. 3.6 which includes:

- Process owner(s).
- Process name (potentially including some description for clarification where required).

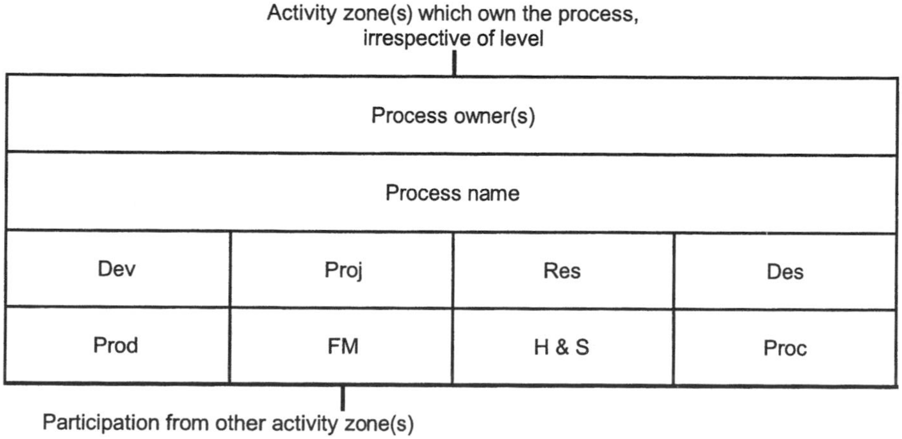

Activity zone(s) which own the process,
irrespective of level

Process owner(s)			
Process name			
Dev	Proj	Res	Des
Prod	FM	H & S	Proc

Participation from other activity zone(s)

Fig. 3.6 Process symbol.

- An indication of likely/potential participation from other activity zones in the process.

Furthermore, inputs and outputs from a process can be shown as illustrated in Fig. 3.7.

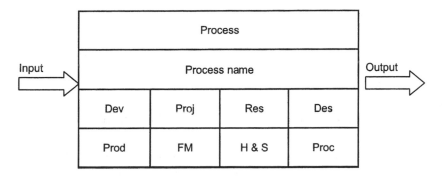

Fig. 3.7 Inputs and outputs to the process.

Inputs

For clarity, inputs to a process are only shown where they form a logical dependency from another process at that level on the same diagram. All other inputs from different phases or activity zones are not shown but are traceable through the modelling database.

Outputs and deliverables

All processes by definition have an output. Some of these can be called 'deliverables' where the information is in a form (or document) that should be named for easy reference and use in other processes. The maps only illustrate the level 1 deliverables for simplicity and space purposes. The outputs from all level 2 and 3 processes will be included in the modelling database.

Process levels

The maps contain three levels that are independent in that there are no interactions between them. These are defined as follows (see Fig. 3.8):

- Level 1 contains high-level processes and their deliverables as identified in the Process Protocol map.

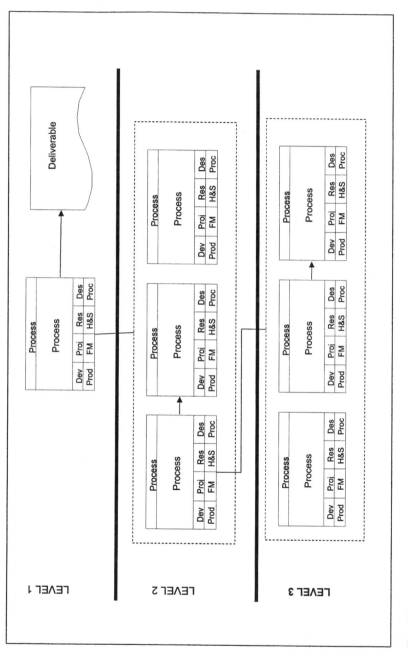

Fig. 3.8   Process level and decomposition.

- Level 2 contains the sub-processes of the main process at level 1 (i.e. what the level 1 process consists of) and how those sub-processes interact with each other (i.e. how the level 1 process is undertaken).
- Level 3 contains the sub-processes of the processes at level 2 (what the level 2 processes consist of) and how those sub-processes interact with each other (how the Level 2 processes are undertaken).

Other attributes related to the process levels include:

- The three levels are separated by black lines.
- A single line connects a process at one level with its group of sub-processes at the level below to denote decomposition as shown in Fig. 3.8.
- Processes can have a logical dependency within a level and this is shown by an arrow as illustrated in Fig. 3.9.
- The participation in a process is shown in the table at the bottom of the process symbol and where such participation does not occur the respective table cell will be knocked back, i.e. appear faded in relation to the other cells.

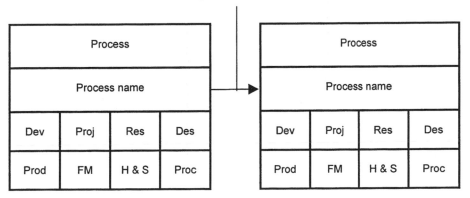

Fig. 3.9   Logical dependency.

Once the decomposition of the Process Protocol down to level 2 and 3 was complete there were over 1000 processes defined and it became obvious that an IT support tool was necessary to support the modification and adaption of the generic process to specific projects. Therefore an IT support tool was developed.

Process support: the Process Protocol toolkit

The Process Protocol toolkit is a software tool to help the construction industry to adopt the Process Protocol and to enable effective project information management and knowledge-sharing between projects based on a consistent framework. This is achieved by demonstrating and communicating the underlying principles and philosophies of the Process Protocol framework. The toolkit is composed of two major components: the *process map creation tool* and the *process management tool*.

Process map creation tool

The process map creation tool is a process-mapping knowledge tool specially designed for the creation of the project process map based on the Process Protocol framework. It automates the map creation process and guides the user who might lack knowledge of the Process Protocol to create a project process map at the early stage of a project. Users will be able to tailor and customise the process map to suit their own project and company requirements while having access to the template and default features that demonstrate the knowledge derived from previous projects and the Process Protocol framework.

To some extent the process map creation tool is similar to many process modelling tools that have been available on the software market for years. Many companies have adopted a process-orientated view of their business operation, replacing the traditional functional viewpoint to achieve a better integration of operation. However, the aim of the process map creation tool is to create not only a process model but the basic information structure derived from the Process Protocol framework. It enables the production of a project process map based on the generic Process Protocol framework and it can be used across different projects. There are three major components in the tool: main creation tool, generic processes data store and project process data store.

The main creation tool provides the functions for data retrieval, map creation and map customisation for different projects. Users will be able to define their processes and create the project process map by referring to the generic processes provided by the Process Protocol. All the generic processes developed are stored in the generic process data store built according to the Process Protocol data model. The project process map created by users is stored in the project process data store which becomes the basis of the process management tool.

Figure 3.10 is a screenshot of the prototype of the process map creation tool. It is a standalone Microsoft Windows application developed using the

Fig. 3.10 Screenshot of process creation toolkit.

Microsoft Visual Basic programming tool. Its interface consist of three main parts.

- *Process tree*

On the left side of the window the process tree is used in a similar Windows file explorer style to show the decomposition structure of the process map. Processes in three different levels are represented in a process tree hierarchy. Processes in the process tree can be selected by mouse click and the corresponding process in the process map will be highlighted. In Fig. 3.10 the process 'Update Financial Factors' is selected and the same process is highlighted in the process map.

- *Process map*

The process map is a visual representation of the Process Protocol map and interacts with the process tree on the left. Processes in different levels are represented in different colours on the screen (Fig. 3.10).

- *Process details*

All the information associated with each process is shown in the process details dialogue box. It includes name, process level, process owner, description, phase and type. Figure 3.10 shows the detailed information of process 'Update Financial Factors'.

Process management tool

The process management tool is a knowledge-based project information management system which integrates the process model as its core information framework. We have identified the requirement of the process management tool which should provide:

- Knowledge capture functionalities such as document/drawing publishing to record project activities and archive project documentation based on the process frame created by the process creation tool.
- Knowledge development functionalities using data mining or OLAP techniques to analyse the project information in order to identify the information pattern, potential conflict such as potential resource issues on site, optimised construction programme and construction process simulation for possible crash detection.
- Knowledge sharing functionalities such as messaging service, email notification and document sharing.
- Knowledge utilisation functionalities such as web-based interface, personalised project information page for each user and fast search tool for document or information retrieval.

In such a knowledge-based environment teams can reduce costs and save time as they gather and disseminate information throughout the project life cycle. Furthermore, the integrated project process map will become the route map to help and guide the project management team to monitor and track project progress, documents and other factors. The centralised project information can be retrieved for future projects where the knowledge captured and stored can be used to inform decision-making and value engineering.

The proposed process management tool has some attributes currently found in some web-based project management systems – also called *project extranet* – which since 2001 have grown rapidly in the UK. Over 1500 projects with a total capital value of more than £20 billion are now managed by web-based project management systems. (See the Appendix for a list of such tools).

The main function of a project extranet is to share project documents and the current systems are now not only allowing the sharing of documents through the network but are also enabling users to view most formats of computer files without installing any extra software. Moreover, users are able to mark up (redline, comment) and make revisions which become a part of the original document. A project extranet is also a restricted network for the project team with every user identified by user ID and password. It is therefore possible to automatically track and log the activities of individual users and see, for example, who made what comment or who viewed or modified a particular file. It is also possible to introduce the project hierarchy and assign different access levels to ensure that information is seen by the right group of users.

Another important advantage of project extranet is that all the project participants have access to the most up-to-date versions of project information. This will, in theory, significantly reduce the expensive mistakes caused by someone working on out-of-date information which are quite common in the construction industry. Furthermore, the cost of sending and printing the project documentation will be reduced since most of the documents are exchanged electronically. However, the current systems are mainly concentrated on storing and indexing the project documents and sharing the documents via the Internet. The major benefit of the proposed process management tool is in using the construction process as a knowledge/information framework to capture project information, making it possible to reuse or analyse such information for future projects.

Summary

The principles of the Process Protocol can be summarised as a model capable of representing the diverse interests of all the parties involved in the process, which is sufficiently repeatable and definable to allow IT to be devised to support its management. The simplicity of the Protocol allows it to be interpreted and applied at a variety of strategic levels across a variety of scales of project using combinations of virtual teams and IT systems, offering clarity in terms of what is required from whom, when and with whose co-operation; for whom the requirements are to be delivered for what purpose; and how they will be evaluated (through the phase review board).

Other underlying principles of the Process Protocol are the standardisation of deliverables and roles associated with achieving, managing and reviewing the process and the product and the introduction of organisational and industry-wide co-ordinating process improvement programmes that incorporate facets of process and IT/IS capability, all of which are based in a philosophy of the early entry of stakeholders and functionaries with an emphasis on the design and planning to minimise error and reviewing during the construction phase.

The Process Protocol is divided into a series of sub-phases defined as pre-project, pre-construction, construction and post-construction and within each of those major phases there are sub-phases which can be operated concurrently or concentrated to make the process more efficient in smaller-scale projects.

Novelty arises within the Process Protocol in a number of areas, in particular:

- The extension of the boundaries of the design and construction process into the requirements capture phase of pre-briefing client decision-making.
- The extension of the boundary of the process beyond practicable completion to allow the management of use and the learning from performance in use to improve the product and process for future projects.
- The creation of an explicit process management and change management role to co-ordinate the functionaries and deliverables associated with the process, the information that supports the functional roles and a stable platform to allow innovations in process and products and operations to be facilitated in a co-ordinated and repeatable manner.

This chapter has by its description of the development of the Process Protocol and its supporting IT given an insight into the work necessary to develop consistent processes for the management of design and construction projects. This work provides a starting point on which to build approaches for the industry using reference to manufacturing. Chapter 4 provides and insight into the issues related to introducing process management into the construction industry.

4

Implementation Issues

'Companies often feel helpless in the face of rapid changes in technology, competitive structure, or market demands. Many look as if they have lost their intensity; their focus, even their will to compete... Redefining the fundamental philosophy and structure can seem like trying to change the course of an ocean liner by swimming alongside and pushing it... Fundamental change can and must be made through a succession of mutually reinforcing measures. And not only the results of change but the process itself can be exhilarating.'

(Deschamps & Ranganath Nayak, 1995)

The introduction of any technique, process or technology into an organisation or an industry requires careful consideration and planning. All too often a 'flavour of the month' management issue is embraced with gusto and implemented by senior management using a management guru or consultancy to impose new ways of working on a reluctant workforce. This often results in failure or marginal take-up. In considering the implementation of a process management approach to design and construction it is important to understand the implementation issues and the barriers that may arise. This chapter uses the knowledge gained through the development of the Process Protocol and, by looking at its implications for a number of construction projects (one is used in this chapter as a case study), highlights the requirements that must be considered before a process management approach is applied to design and construction. The chapter goes on to discuss how a process map such as the Process Protocol can be used for the integration of additional issues facing the industry such as sustainability, standardisation and modularisation (see Fig. 4.1).

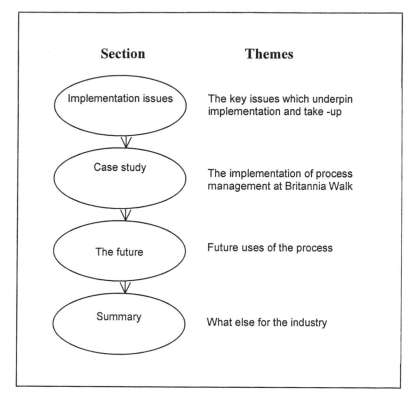

Section   **Themes**

Implementation issues   The key issues which underpin
                            implementation and take -up

Case study   The implementation of process
                            management at Britannia Walk

The future   Future uses of the process

Summary   What else for the industry

Fig. 4.1   Chapter map.

Implementation issues

Culture

As with much of the research in the construction industry today, some of
the problem factors which were identified during the development of the
Process Protocol especially in the pre-project phase of existing design and
construction processes were issues relating to the culture of the construction industry and the interpretations of this culture by the client body:
that there is a culture which is not homogenous across the entire
industry, and that there are sub-cultures and organisation cultures within
a wider industry which is in general slow, or indeed resistant, to change.
This is also a culture which, post-Latham, many of the industry's leading
bodies were working very hard to try to change and in the mid-nineties
there did indeed appear to be a desire for change towards a more co-
operative, less confrontational culture, hence the introduction of partnering relationships. This was seen as being a potential advantage for the
introduction of a unifying Process Protocol. Indeed there needs to be both

a push and a pull, from within an organisation and within a project, for a process to be adopted into a culture. Understanding and recognising the need for a culture change is crucial to the implementation and adoption of new process approaches.

Application

Implementation of a process such as the Process Protocol must recognise the need for adaptation to particular circumstances. Indeed it was also appreciated that there was no single best way to control all design and construction circumstances and that a generic protocol would not be universally applicable or universally applied. Rather, as a high-level process, it was intended that a generic and adaptable set of principles would be used and which would be designed for consistent application in a repeatable form but with the scope for attunement for the specific detailed circumstances of individual projects, teams and client needs.

Early involvement

There is clearly a need for an earlier and fuller involvement of design construction professionals in the pre-project advisory stage where clients would be deciding upon construction as one of the many possible options to improve their facilities. It was articulated by the participating industrialists and the clients in the Process Protocol project that a lack of consultation opportunity between the industry and client bodies, in part because of the contractual conventions under which the industry was normally appointed, was leading to a chronic under-provision of professional advice and transfer of previous experiences to new projects, even with serial clients operating in a non-partnering arrangement.

Strategic or term partnering arrangements which span across a series of projects offer scope to allow the early pre-project phases to be addressed in a more positive manner and in particular to build upon the experiences of the process and the product from previous projects. This enables the client, contractor and designer to all move forward together to a better quality product and a better quality process. In such circumstances a Process Protocol which allows or facilitates the development of virtual teams and the co-ordination of a series of investigative activities in the pre-project phase is a potential vehicle for achieving early requirements capture of the client's needs. This also creates the operational platform for the introduction of early IT support at the briefing stage.

Team role

The industry has generally been built around functions and organisations working together. It is recognised that these must become fundamentally more effective. Obviously projects require virtual teams and should operate around a team-building philosophy in order to establish team relationships at an early stage both at and between managerial and operational levels. It is also preferential that these teams, or the spirit of co-operation between the teams, and a protocol for their co-operation should extend beyond individual projects.

One of the characteristics of the industry is the creation and dissolution of teams between projects and although term partnering is increasing, the stability of team structure within organisations and between co-operating organisations there is still a great likelihood that individual projects will have a degree of diversity of team membership and team operation. It is here that one of the greatest potentials of introducing a generic Process Protocol arises – specifically, to allow a consistent set of processes to be adopted and applied in a flexible manner across a series of projects by all professionals involved in those projects and thereby to enable them to carry forward a set of implicit and explicit techniques and managerial approaches to the design and construction process. It is clear that in order to gain the consistency and knowledge transfer benefits of introducing a process management approach, the creation and management of teams must also be given considerable attention.

Contractual changes and feasibility

Obviously if professionals, contractors and sub-contractors are to contribute to the very early stages in a project there is an important need for changes in contractual arrangements, possibly to include retention-based approaches or consultative-fee-based arrangements for pre-project consultation from a variety of professionals who can improve the requirements capture phase for the clients and allow them to make their early stage decisions on construction or alternative project solutions from a base of informed knowledge rather than inexperienced ignorance.

The issue of copyright over ideas and transfer of copyright of ideas, between products and processes from individual projects generally or from project to project, also needs to be considered before this system can practicably be operated. These sorts of changes in contractual and consultative arrangement are most likely to be facilitated through implicit or, particularly, explicit strategic (term) partnering arrangements between clients and contractors, and also between contractors and sub-contractors within the industry or even partnering within the supply chain.

Risk allocation

The allocation of risk is an issue with which the industry is currently struggling, especially with regard to PFI and similar types of projects. It is most definitely important to tackle the identification and allocation of risk when implementing a generic process, especially at an early pre-project phase. It was seen by all industry partners as being important that within the protocol as a whole there would be a clear sub-protocol for the early addressing of risk and risk allocation by the stakeholders involved in the pre-project process phases. For this to be meaningful across the design, construction and use phases it was essential that all the stakeholders involved in the process of creating the building and using it would be involved in this aspect. Risk management should also be implemented throughout the process and risk allocated within the framework.

Process alignment

There are often a range of processes that operate within individual professions and organisations. This is common among client organisations that use process management in the course of their own business and transfer that knowledge to managing construction projects. While these client-led process protocols such as the BAA project process provide stability within an individual client base and, if spread across a range of clients, may produce a relative stability and predictability across projects within a certain sector of the industry, this in itself will not ensure generality and consistency of a process across a whole industry.

Clearly the range of processes being developed already and the scope of existing IT support and other process application tools mean that the journey towards a consistent repeatable process is one with which a generic Process Protocol may assist but to which it is unlikely to provide the final solution. What is essential is that the system should be integrated and should embody a series of radical changes to the existing processes, the sum collection of which will be identifiable by the industry as an improvement in process and product and will in the short term produce measurable benefits in terms of time, cost and quality. Therefore the introduction of a process approach must be evaluated from the onset.

Co-ordination

The journey from an existing, fragmented, unco-ordinated and undefinable process requires a significant degree of co-ordination. This co-ordination needs to involve data exchange management, information management, professional integration management and the effective application of change management techniques within individual organisations and across organisations.

It is clear from the studies that an explicit information management protocol requires to be linked with a consistent Process Protocol and that, while this may be supported by an information management system in the foreseeable future, at the moment, it will essentially be a paper-based exercise involving the co-ordination of various professional and process activities which themselves may be supported by individual or partially integrated tools, information technology or information systems. For either a partial or full application of a Process Protocol to occur successfully there have to be both clearly defined deliverables that are associated with certain functions and activities associated with achieving particular deliverables.

Process complexity

The Process Protocol project identified a series of ten phases which fell into the pre-project, pre-construction, construction and post-construction phases of the design and construction process. For complex projects it may be necessary for the project to be explicitly directed through all of the discrete process phases. However, for less complex projects and for projects which involve a degree of repetition from previous projects (and can therefore adopt elements of previous process and product practices which are being captured through a systematic collection of legacy information) it may be possible to collapse some of the phases using the soft gate concept to allow the process to be managed through less explicit phase review meetings. This flexibility is akin to concurrency and if managed consistently by a process management role can allow the flexible application of the processes in accordance with the existing experience, size of project, client needs and consistency of team membership in relation to previous projects. The implementation of process requires application in a flexible manner and, as such, it is most appropriate to start with high-level strategies process definition and implementation before the application of more detailed operational processes.

Activity predominance

To avoid organisational and company orientation of the process and to ensure an industry view it was considered essential that the process phases in the Process Protocol should be activity driven and based on deliverables which were identifiable according to phase activity team and professional involvement; and that these deliverables should also be driven by the functions undertaking the activity or task. It was important that the processes were transparent and defined simply and also could be related to a simple straightforward icon to which people could attach their interpretations of it. This should extend to the activities of the various individual functionaries involved in process and the deliverables that they have undertaken to achieve (including the timing of these).

Design fixity

The principle of design fixity is contentious and evidence from other sectors and other applications indicates that a range of approaches to design fixity is applied depending on the type of product and the needs of the client. Obviously from a construction perspective complete design fixity prior to construction, and preferably prior to pre-construction planning, would allow a certainty over the construction process. This in itself could produce dramatic improvements in time, cost and quality of the finalised product. It could also support a reduction in construction-based errors, particularly those occurring at the interfaces between different products and trade activities. However, in the context of most projects which are being built to an explicit time limit, the stipulation that all design and planning must be completed before construction starts is unrealistic and probably ultimately more expensive in terms of time and perhaps cost on a time and design critical project.

The principle of design fixity within a process is that the design elements are signed off once agreed; that this should be done as early as possible within the conceptual and full design and pre-planning phases prior to construction; and that in all likelihood they will be phased in according to the order of design and/or construction. For instance, it would be a desirable goal for structural frame detailing to be fixed as early as possible to allow the service planning to be carried forward in the knowledge that there will be no or few changes. However, even here there may be a possibility within the contractual applications associated with the process for clauses relating to changes late in the process which would stipulate that changes beyond the agreed design fixed stage would be allowed but the cost of disruption would be borne by the client rather than the contractor. The detail of fix schedule and scope needs to be decided and managed by the phase review/project board and implemented by the process/project manager.

Process/change management

The introduction of a consistent process requires management and superficially this will add cost to the existing process. However, the net gain associated with having a more systematically and systemically designed and managed process would probably outweigh the cost of administrating it. The process management role is multi-faceted and there is no existing individual professional identity suitable for undertaking this role. Rather, an individual with good people-management skills and familiarity with the range of professional requirements within processes would be best qualified to undertake it. This role could come from any or many of the professions.

Another aspect of the process management role is that of change management. In smaller projects it is probable that, on a cost basis and with the scale of work involved, process management and change management (change management defined in terms of information change rather than organisational change) could be undertaken by the same individual. This would clearly have benefits in terms of avoiding communications between different individuals undertaking these complementary roles but in larger projects the scale of work involved would probably require different individuals to undertake the two roles. In particular, the change management role requires development, implementation and management of communication protocols across the whole series of process phases. If these protocols and their operation could be devised in a generic manner the facilitating role of the change manager could be integrated relatively easily, with improvements in IT and IS, into organisations taking up the process. There will clearly be a cost associated with operating the change management role but this may be fairly easy to demonstrate as a benefit in relation to the potential confusion that can occur with the errors in information communication that are widely acknowledged as besetting the industry processes currently in operation. The change management role allows process information to be used more strategically not only within the individual process or project but also, in the long term and more importantly, between processes, between projects and across the industry horizontally and vertically.

Phase reviews

In order to ensure that the implementation of the process works for the benefit of the project and the various stakeholders, phase reviews should be used throughout a process. A phase review board comprised of stakeholders who have early process entry and represent the virtual company involved in the development, creation, design and construction of the project should be assembled. The role of the phase review board is to assess phase progress and to adjudicate on the continuation of the process and the project. Clearly this brings together client, designer and contractor in a co-ordinated manner. However, it also raises the opportunity of introducing facility management and individual professional expertise at the early stages of feasibility and design study. The role of the phase review board includes the review of the product and the process during design and construction and the review of finances and strategic planning in accordance with the overall business objectives for which construction is usually a secondary support initiative; and, in particular, it allows all stakeholders to understand the existing and potential problems which may arise during the process especially at an early a stage when changes can be made with the least disruption.

The process manager should review the process, its management and the management of deliverables from the various functional groups in order to provide the phase rcvicw board with the necessary information to make its decisions at each of the meetings associated with soft and hard gates in the process. For this to operate in a repeatable form it would be essential that the professional roles and inter-relationships within the design and construction process be defined. The interpretation of the functional role, the deliverables and the activities and distribution of these workloads between professionals within the various functional teams should be decided by the process manager flexibly and in accordance with the specific project needs and the capabilities of the team which is brought together.

IT and process inter-relationship

There is a clear relationship between sophisticated process capability within organisations or virtual teams operating projects and the information technology that they intend to utilise. Where there are low levels of organisational IT preparedness the efficiency and effectiveness of the processes can be impaired. Indeed even more crucial is where there are differences within and between organisations in terms of capability and IT application and inconsistency between process and IT protocols. The confusion that this creates in terms of data and information exchange can be very damaging to process efficiency. In the long term initiatives such as those by the International Association for Interoperability to integrate the protocols for the use and application of information technology-based knowledge and processes will support smoother integration and application. In the immediate future it is essential that compatibility and IT integration are managed at the onset of the project.

Business implementation of IT

It could be argued that the costs in terms of purchasing and developing the software, training personnel to use the software, management and maintenance of the applications and associated data, communications links and so forth, could far outweigh any cost benefits of implementing such a wide-ranging system. Also industry-wide reluctance may prevent the adoption of an integrated system because of threats to various professional practices such as professional property developers who would normally undertake economic appraisal at the early stages of a project and quantity surveyors who might have a reduced involvement during the whole project.

The ownership of the information produced could be a bone of contention that might hinder the implementation of such an overall system. For example, the contractor might want all of the information produced to be incorporated into their archive while the client would possibly expect all of the project information to be theirs and copyrighted as they had paid for the building. With regard to the compatibility of the IT with the Process Protocol or indeed any holistic process, it could be argued that with the speed of development of a project design when utilising an integrated IT map there might be substantial backlogs and idle time for various project participants. This could occur because of the various bureaucratic stages and processes of the Process Protocol slowing down the information transfer from process to process and stage to stage. However, these issues should not be considered barriers to planning the integrated implementation of process and IT. Rather they should provide insights into the planning and management of such systems.

A case study of implementation: Britannia Walk

The background

In order to understand some of the issues in process development and implementation the Process Protocol team (Chapter 3) set up a number of test case studies. The Britannia Walk project was one such case.

The Britannia study was led by the University of Salford and Alfred McAlpine. It followed their previous collaboration on the developmental research that resulted in the Process Protocol. The study was of the use of the Process Protocol and its key principles on Britannia Walk. The Britannia Walk project is a complex brownfield development situated in Hackney, London, and occupies a site close to Moorfield's Eye Hospital which had for many years been under-utilised as an NCP car park. The development, which had been in gestation since 1996, was due for completion in 2002. The case study was undertaken between 1999 and 2002. The project was a complex one, mostly because of the large number of stakeholders and the mixed-use nature of the scheme with different types of accommodation (ranging from a pharmaceutical manufacturing unit to key worker accommodation and speculative offices); also the funding of the project, which consisted of public and private finance, and the high design standards of the scheme. All of this meant that from the outset the project would be a challenge to manage effectively.

The research

The research and implementation (R&I) project began in October 1999. The first stage was to identify the existing process that had been established on the project so far and ascertain how this was being managed. The research tracked the project and identified how each of the deliverables, as recommended by the Process Protocol, had been implemented. The research then implemented aspects of the Protocol that were not being effectively undertaken on the project and assessed their impact.

This first stage of the R&I project revealed that although the Protocol deliverables were not being documented, many of their underlying components were being addressed whether through negotiation during the early formative stages of the project or by being contained in other key project documentation such as planning submissions, selection documents, employer's requirements and user briefs. These components were documented more rigorously as the project progressed. This resulted in the project process map.

This tracking exercise revealed that although many of the deliverables were addressed, the mechanisms for managing the process, such as stage reviews, clearly defined deliverables, open communication between participants and establishing a common project focus, were not all rigorously implemented. The implementation aspects of the R&I project attempted to address these problems. There was also a need for the principles of the protocol to be tailored to help manage more specific operational tasks. The principles of the Protocol were used to produce more detailed sub-processes (represented by process maps) for the design and production stages of the project. These tools proved to be of considerable use and, according to the team, provided a fresh approach to solving common problems. The R&I project also revealed how important team-building and partnering principles are in establishing, agreeing and adhering to a predefined process strategy.

General results

The general conclusion drawn from the work was that if a process strategy is established early in a project which defines terminology, deliverables and stage-gates and utilises operational tools similar to those developed on this R&I project then the task of managing the high-level project process and its constituent operational sub-processes is made much easier.

The team were also of the opinion that the key elements of the Protocol being the common project view, clearly defined team deliverables, stage-gates and process maps would lead to a cohesive project team who are clear on what they have to deliver and by when.

The work revealed that when the project was encountering problems aspects of the protocol methodology were missing and when these aspects were introduced the team saw the benefits. The research concluded that project performance could be substantially improved if such principles were introduced early enough in the project as part of a pre-defined project-specific process strategy as recommended by the Process Protocol methodology

The Project Process map in (Fig. 4.2) illustrates the process that was employed for the project. As can be seen, the processes and deliverables recommended by the protocol were largely employed on the project although some were more rigorously implemented than others. The implementation (action) research began on the project in phase 5. The project tracking that took place in the first stage of the research revealed that many aspects of the key deliverables for the early stages of the project that were recommended by the protocol were considered even if a corresponding document was not produced. More formal documentation was produced at phase 3 when the employer's agent was appointed for the project and began to produce the contractual documents required for the project to progress. The initial planning submission was undertaken in phase 3 followed by the preparation of the partner selection document which was produced in phase 4.

The project also accelerated more rapidly after the selection of the partner contractor in phase 5 who undertook a principal process management role regarding the integration of design, procurement and production elements. One of the first key activities that the contractor undertook was the value engineering report and risk assessment exercise which resulted in several million pounds being trimmed off the project budget. During phase 5 the first draft of the key ER (employer's requirements) document was produced by the employer's agent. This key document encapsulated many aspects of the Process Protocol deliverables (such as components of the business case and project brief). Although in general the critical factors for effective process management were in place there were key elements that were not

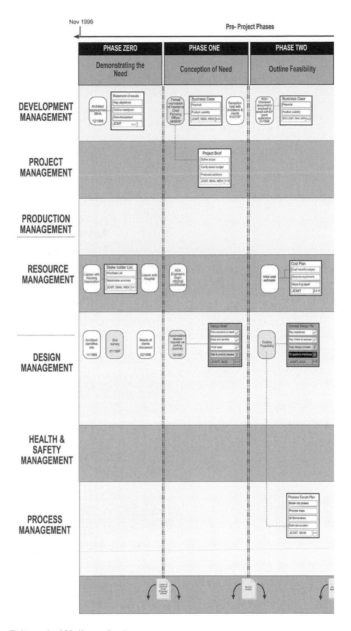

Fig. 4.2   Britannia Walk project process map.

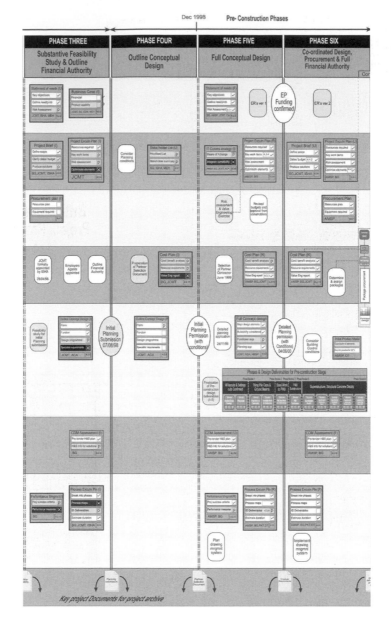

Fig. 4.2 *Continued.*

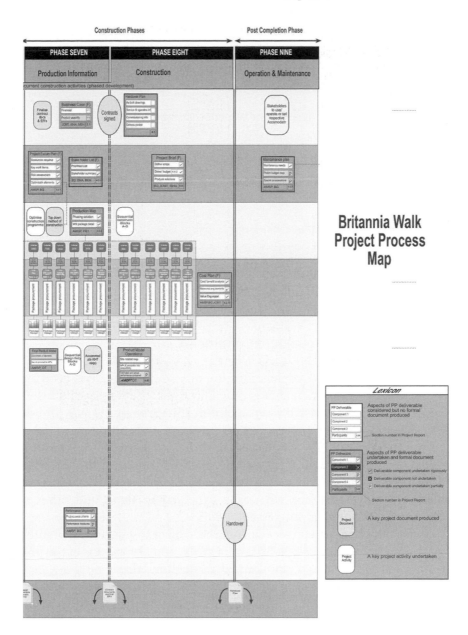

Fig. 4.2   *Continued.*

undertaken on the project. These could have made a considerable improvement to the progress of the project if they had been planned early on and implemented by a team working to a pre-agreed process strategy. These key elements are as follows:

- Early planning and agreement of process strategy.
- Setting terminology.
- Buy-in from the team.
- Communication.
- Standard deliverable sets.
- Partnering.
- Project workshop.
- A driven process.
- The right level of detail.

These will be briefly discussed below.

Early planning and agreement of process strategy

Front-end planning and implementation of a strategy that has been agreed by the key project participants is crucial. As the clients stated on the project, a structured process is very useful in the early stages as it imposes a clear direction for the project before contracts and more formal relationships are in place. This is especially true for clients who have little experience of construction, as was the case with one who stated that the Protocol would have been of enormous use to them at the beginning of the project to help them see what was expected of them and other team members as the project progressed.

The initial stages of a project can also be subject to a great deal of informal negotiation as was the case at Britannia Walk, and key aspects of the process strategy were considered and defined throughout these early stages. It is important to be able to document and record these as prescribed by the protocol so that an archive exists throughout the duration of the project and enables information to be easily reviewed and updated.

Set the terminology

It is also important for a process strategy to be implemented on a project prior to any contract documentation being in place. Such documentation (for example, the ER or partner selection document) defines the terminology for the project. If a process strategy is imposed with its own definitions for deliverables and procedures after a contract document has begun to define a different terminology confusion can and will occur.

Buy-in from the team

Even if a clear and well-structured process strategy is defined for a project it will only be effectively implemented if the team support it and adhere to it. Involving the team when specifying project deliverables is important in obtaining their buy-in and ensuring that the deliverables and stage-gates are effectively planned for the project. It is the team who must provide the information and satisfy the deliverables and they should therefore be consulted regarding when they can provide them by.

Communication

Communication of strategy

The process strategy needs to be communicated to the whole team to avoid marginalising disciplines and to keep teams informed of progress. Mapping the process was shown to be of great use in communicating it to project participants and in providing a common project view for the entire team. To maintain such a common focus on the project the strategy (preferably represented by a process map) needs to be used as a tool for all participants and should not simply reside in the project manager's office.

Team communication

It is also vital that interdisciplinary communication is well organised. If milestones and stage-gates along the defined process are continually missed because of inadequate communication of information between participants at an operational level then the process approach can lose credibility, making it more difficult to get members to conform to its principles. The project communication strategy is therefore vital to project success and needs to be defined and implemented up front when the overall process is being mapped for the project. This is equally true of paper-based and IT-driven communication strategies.

On the Britannia Walk project the communication strategy had some problems during the concept and detailed design phases even though the managers were rigorous in producing design programmes and information required schedules. Fixity could not be achieved when scheduled because of the inherent problems in the team obtaining information from other team members at the right time. In essence the managers were adopting principles of the protocol, i.e. programming, defining the design process and stipulating the design deliverables, but could not hit the desired deadlines because the team were not exchanging information in a structured and organised way. This changed to some extent with the introduction of the drawing management system which enabled participants to access a central archive of drawing information and easily identify what stage of completion specific drawings were at. The system improved communication and management of the drawings to a significant degree. The communication strategy was further enhanced by establishing team deliverables (as opposed to individual ones) for work package specific design information thereby forcing participants to work together more closely and share information more openly.

Standard deliverable sets

The structuring of project information into clearly defined team deliverables means that constituent information will come from a range of different team members and organisational types. It is important to have a standard method of representing the information so that it can be reviewed easily. On the Britannia Walk project the engineers adopted the structure and terminology used for the design deliverables when reporting progress. According to the project manager this made a significant difference, allowing him to monitor progress more speedily. However, other participants did not do the same. Performance could have been improved considerably if all had adopted a standard method and format when organising and exchanging information.

Partnering

The Process Protocol does not focus on partnering aspects but they are helpful as a mechanism to support the process approach. The principles of a formal partnering arrangement exist to help establish a cohesive team (e.g. team buy-in, effective communication). This aspect was lacking on the Britannia Walk project and the effectiveness with which process strategy could be implemented suffered as a direct result. Some team members reported that although there was talk of partnering nothing was really undertaken after the partner selection document was produced.

Project workshop

The principles embodied by the Protocol were to some project participants quite unfamiliar and it was important for the project team to understand the approach. With the introduction of any new management method the team who will be working within the new framework need to be informed of the approach and have an opportunity to contribute to any strategy development. Some form of project workshop is therefore recommended at least with principal members of the client, management and consultant teams. It was intended to hold a workshop for Britannia Walk towards the end of phase 5. The aim was to include the key disciplines and clients, the newly appointed design consultants and the 'partnering' sub-contractors. The aim was to get the team together for two days and one night (during the week, not at the weekend) and undertake a mixture of seminars and team-building exercises. It was to be held at an outward bound centre and the focus was to be on having fun, the team getting to know each other and defining themselves and agreeing as a team, key partnering objectives and an initial process framework.

Unfortunately this did not occur. It is the opinion of the research team and project manager that such a workshop would have had a significant impact on the project as it would have performed a dual role. It would have enabled a more refined partnering approach to be formalised and would have provided an opportunity for participants to provide their comments on the Process Protocol, enabling them to contribute to strategy development by defining how they would like their particular operational sub-processes to be defined. It would also have enabled some new terminology regarding protocol deliverables or project phases to be established on the project and consequently would have enabled the process strategy to be defined and implemented more easily.

A driven process

The role of the process manager became clear during the course of the Britannia Walk project. At first the client's representative and contractor's project manager were saying that they believed the process manager's role would be undertaken by the project manager, architect and client's representative. However, it became clear as the project progressed that the process approach was required in order to help guide the managers in the use of process management methods and also to keep a 'helicopter view' on the project. The contractor's management team were for obvious reasons highly focused on procurement and production activities even though they were also managing the design process. The client's representative was more focused on the contracts and development management activities and there was a lack of co-ordination between design, procurement and production. The role of the process manager is to consider all things equally from an independent perspective and the importance of this role became clear. The contractor's project manager took on such a role for phase 6 of the project when the production map was produced. Although still interested in receiving information from the team at the right time to enable timely procurement and production activities, the identification of the needs of other disciplines, the use of team not individual deliverables, more open information sharing and improved co-ordination of activities all helped the project manager to orchestrate the various elements collectively and to manage the process for these stages, not simply to manage the construction.

On Britannia Walk the role of process manager was identified as important, with the project manager stating that he thought the role of a process manager was required and that he couldn't have undertaken both roles. In time he believed there could be more crossover but only once project managers had been trained and were suitably knowledgeable regarding process management, and if there were no hidden conflicts of interest.

The right level of detail

When the Process Protocol guidelines were provided to the team initially there were various options regarding the document and the approach. As has been said, the clients were of the opinion that the document was very useful in providing clarity regarding the intended process and enabling them to be 'better informed and prepared for upcoming stages and events in the project'. The practitioners, however, required more detail and there was concern that the approach was too high-level. The general census of opinion was that the protocol approach was highly valid and the team liked the idea of stage-gates, the structuring of required information into team deliverables and so forth, yet they wanted to know how this was put into practice at an operational level.

The research team therefore sought to develop operational aspects of the approach and what was learnt from this was the importance of ensuring that team members know how they are fitting into the overall project process.

There also needs to be a set of working procedures which are designed to help provide information in the right format and ensure timely delivery. The method of structuring specific design review meetings, for example, which were related to specific activities and deliverables represents the type of procedure that should be in place in any process management strategy to ensure that information is communicated and managed in a compatible and integrated way between different facets of the project organisation.

There are far more participants working on a project at an operational level than there are managing it at an executive level and their involvement is therefore crucial to the success of the project. Although there are no defined operational procedures as yet prescribed by the Process Protocol this needs to be considered up front when planning any process strategy. The aim should not be to radically change the way people work but to ensure that the process management approach extends to include all participants, not simply the project managers. However, developing the operational process to be used will require commitment from the team and buy-in of the defined processes.

Summary of the case

The Process Protocol research and implementation work undertaken on the Britannia Walk project has afforded a valuable insight into how the current approach of a highly experienced project team differs from the recommended approach defined in the Process Protocol. It revealed that at a high level the two approaches are similar, with many Process Protocol deliverables being in tune with key project activities. However, the work has also illustrated that at a more operational level the principles of effective process management are more difficult to maintain especially if the team has not been set up as a cohesive unit working to an agreed strategy. Through implementation the project has revealed that by mapping together processes that are normally considered in isolation, so that participants are aware of how they are contributing, a far more streamlined process can be established for each work package. The efficiency of these operational sub-processes is crucial in maintaining the higher-level strategy developed at the beginning of the project.

The project also identified the importance of establishing an effective communication strategy and fostering principles of collaboration and open-book information-sharing to avoid the common problems associated with differing incentives and responsibilities between design and production disciplines.

The operational sub-processes on the Britannia Walk project were developed from the principles of the Process Protocol (i.e. to achieve fixity, to visualise strategy and to integrate interdisciplinary teams and their respective activities and deliverables) and their development has demonstrated that these key principles work in practice. The lack of opportunity to establish the process strategy up front on the project meant that the process approach as recommended by the Protocol could not be fully implemented. However, even when implemented without front-end planning the Protocol approach dealt effectively with specific problems encountered and was considered by the team to be very beneficial.

In conclusion, this case study illustrated that the Process Protocol, its framework and key principles are workable in laying down the necessary disciplines for a well-managed construction project. It is important, however, to apply the management and implementation of the process from the onset of the project.

The future

It is probable that the introduction of a process approach into the construction industry will be a catalyst for change and will itself need to change as the industry adopts it. The evidence from the case study is that change management and process management are seen as some of the key process changes that need to be put into place first; and that co-ordination of information technology or information systems is another essential requirement. These activities will allow processes to be controlled in a consistent and predictable manner that will allow the maximum product innovation and process innovation.

It is considered to be an essential that the efforts of any process team are loaded towards planning effort, the pre-project and the pre-construction phases of the process on the assumption that this will allow the minimisation of ambiguity and late-stage design changes and thereby allow a greater pre-production planning efficiency and a smoother and more controllable production phase to be achieved. This will also allow the greater predictability associated with this for production and the more rapid and cost-effective introduction of standardisation and prefabrication, the current problems of which tend to occur at the interfaces between trades and elements where pre-planning effort is required to make improvements.

The use of a consistent process is already seen as having spin-off benefits. Critical issues relating to the development of the industry can be applied to the framework, for instance CIRIA have produced guidelines for standardisation, customisation and modularisation. Using the Process Protocol framework, they have alerted the industry to the points in the process that require decisions on such issues to be made and implemented (Fig. 4.3).

Further examples of the use of the process as a framework come from the application of sustainability issues along the process. For instance, those organisations who are concerned with aspects of ecological sustainability can flag these up in the process and identify which activities are responsible for considering such issues.

Conclusion

This chapter has discussed what is necessary for the implementation of a process management approach to design and construction. It has, through

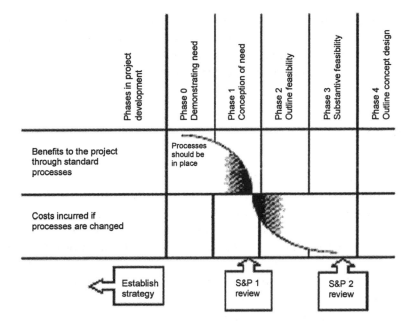

Fig. 4.3 CIRIA guide.

the use of a case study, identified where benefits lie but also what are essential aspects of the implementation process. There is obviously much potential upon which to build such as using the process framework to embrace more widely the basic principles of niche marketing and product platform building. We will see the use of design and construction processes tailored to specific product types such as hospitals and schools and also linking into business processes and, more widely, urban design processes.

The future is about understanding what process you are using and how it relates to the other processes around you, using information technology to connect to those processes and improve the overarching decision-making for the benefit of the individual, the organisation, the project and society as a whole and as a means of communicating critical improvement issues for the industry or for a project.

Appendix
The Process Protocol Phases and Activity Zones Demystified

PHASES

Phase 0: demonstrating the need

What is the problem?

It is important to establish and demonstrate the client's business needs and ensure that problems are defined in detail. Identifying the key stakeholders and their requirements will enable the development of the business case as part of the client's overall business objectives.

Before the phase

The 'user', i.e. business, customer, communicates the problem to the client.

- A master plan (of the client's strategic issues) should be available.

During the phase

- Bring together the business case, facilities management (clients and users).
- Carry out the necessary activities to produce the deliverables.

Goals

- Establish the need for a project to satisfy the client's business requirements.
- Gain approval to proceed to phase 1.

Gate status

- 'Soft' gate. (Note: a 'stakeholder' is defined as 'any person or group who has an interest in the provision or use of the required product.')

Prime activity zone responsibility	Potential activity zone membership
• Development management • Project management • Facilities management • Resource management • Process management	• Development management • Project management • Resource management • Design management • Production management • Facilities management • Health and safety, statutory and legal management • Process management • Change management

Phase 1: conception of need

What are the options and how will they be addressed?

The initial statement of need becomes increasingly defined and developed into a structured brief. To this end all the project stakeholders need to be identified and their requirements captured. Based on those, the purpose of this phase is to answer the question 'What are the options and how will they be addressed?'

Before the phase

- Approval to proceed obtained.
- Approval for funding obtained (probably up to phase 3 depending on the size of the project).
- Results of studies to define need(s) are available.
- Initial stakeholders are identified.

During the phase

- Identify and refine the statement of need(s).
- Develop the project brief according to the business case developed in phase 0.
- Update stakeholder list/group membership.

- Identify options, i.e. do nothing, manage the problem, develop a solution.
- Process execution plan (updated): plan phase review until phase 3.

Goals

- Identify potential solutions to the need and plan for feasibility (phase 2).
- Gain authority and financial approval to proceed to phase 2.

Gate status

- 'Soft' gate.

Prime activity zone responsibility	Potential activity zone membership
• Development management • Project management • Facilities management • Resource management • Process management • Design management	• Development management • Project management • Resource management • Design management • Production management • Facilities management • Health and safety, statutory and legal management • Process management • Change management

Phase 2: outline feasibility

Which option(s) should we consider further?

Many options could be presented as possible solutions to the identified problem. The purpose of this phase is to examine the feasibility of the project and narrow down the solutions that should be considered further. These solutions should offer the best match with the client's objectives and business needs.

Before the phase

- Facilitate the introduction of new project participants.
- Appoint the 'core teams' that will form the activity zones.

During the phase

- Undertake feasibility studies for all options including necessary planning approvals.
- Revise business case.

Goals

- Examine the feasibility of the options presented in phase 1 and decide which ones should be considered for substantive feasibility.
- Gain approval to proceed to phase 3 (substantive feasibility study and outline financial authority).

Gate status

- 'Soft' gate.

Prime activity zone responsibility	Potential activity zone membership
<ul><li>Development management</li><li>Project management</li><li>Resource management</li><li>Design management</li><li>Facilities management</li><li>Process management</li></ul>	<ul><li>Development management</li><li>Project management</li><li>Resource management</li><li>Design management</li><li>Production management</li><li>Facilities management</li><li>Health and safety, statutory and legal management</li><li>Process management</li><li>Change management</li></ul>

Phase 3: substantive feasibility study and outline financial authority

Should the proposed solution(s) be financed for development?

The decision to develop a solution or solutions further will need to be informed by the results of the substantive feasibility study or studies. The purpose of this phase is to finance the 'right' solution for concept design development and outline planning approval.

Before the phase

- Redefine the project brief/business case and project objectives based on outline feasibility results.

- As the options become more defined, consider project success criteria and performance measures.

During the phase

- Challenge the need(s)/opportunities.
- Conduct substantive cost/benefit analyses.
- Submit application(s) for statutory approval(s).
- Produce the concept design plan.

Goals

- Gain approval to proceed to phase 4.
- Gain financial approval (perhaps until phase 5).

Gate status

- 'Hard' gate.

Prime activity zone responsibility	Potential activity zone membership
• Development management • Project management • Resource management • Design management • Facilities management • Health and safety, statutory and legal management • Process management	• Development management • Project management • Resource management • Design management • Production management • Facilities management • Health and safety, statutory and legal management • Process management • Change management

Phase 4: outline conceptual design

How does the solution translate to an outline design?

The purpose of this phase is to translate the chosen option into an outline design solution according to the project brief. A number of potential design solutions are identified and presented for selection. Some of the major design elements should be identified.

Before the phase

- Define the systems, i.e. sub-assemblies.
- Define the criteria for evaluating the systems, e.g. production timescale, cost, resources required, etc.
- Identify major system interfaces and interactions to enable communications and facilitate the introduction of project design teams.
- Facilitate the introduction of key system suppliers.

During the phase

- Iterative development of outline concept design.
- Refine project/system solutions.
- Develop basic schematics, i.e. plans, elevations, etc.
- Identify the implications of system solutions in relation to other system solutions and to the overall project.
- Identify production supply chain.

Goals

- Identify major design elements based on the options presented.
- Gain approval to proceed to phase 5.

Gate status

- 'Soft' gate.

Prime activity zone responsibility	Potential activity zone membership
<ul><li>Development management</li><li>Project management</li><li>Resource management</li><li>Design management</li><li>Production management</li><li>Facilities management</li><li>Health and safety, statutory and legal management</li><li>Process management</li></ul>	<ul><li>Development management</li><li>Project management</li><li>Resource management</li><li>Design management</li><li>Production management</li><li>Facilities management</li><li>Health and safety, statutory and legal management</li><li>Process management</li><li>Change management</li></ul>

Phase 5: full conceptual design

Can we apply for planning permission?

The conceptual design should present the chosen solution in more detailed form to include M&E, architecture, etc. A number of buildability and design studies might be produced to prepare the design for detailed planning approval.

Before the phase

- Review membership of design teams.
- Review evaluation criteria for concept design.
- Identify some of the major systems.

During the phase

- Develop system concept design.
- System interface studies.
- Identify resourcing requirements.

Goals

- Conceptual design and all deliverables ready for detailed planning approval.
- Gain approval to proceed to phase 6.

Gate status

- 'Hard' gate.

Prime activity zone responsibility	Potential activity zone membership
• Development management • Project management • Resource management • Design management • Production management • Facilities management • Health and safety, statutory and legal management • Process management	• Development management • Project management • Resource management • Design management • Production management • Facilities management • Health and safety, statutory and legal management • Process management • Change management

Phase 6: production design, procurement and full financial authority

Are the major design elements fixed?

The purpose of this phase is to ensure the co-ordination of the design information. The detailed information provided should enable the predictability of cost, design, production and maintenance issues among others. Full financial authority will ensure the enactment of production and construction works.

Before the phase

- Review membership of design teams.
- Review evaluation criteria for co-ordinated design.
- Fix major building elements.

During the phase

- Assemble the co-ordinated product model.
- Review and update major deliverables.
- Review supply chain analysis.

Goals

- Fix all major design elements to allow the project to proceed to phase 7.
- Gain approval to proceed to phase 7 and (in most cases) through to the end of the project.
- Gain full financial approval for the project.

Gate status

- 'Hard' gate.

Prime activity zone responsibility	Potential activity zone membership
• Development management • Project management • Resource management • Design management • Production management • Facilities management • Health and safety, statutory and legal management • Process management	• Development management • Project management • Resource management • Design management • Production management • Facilities management • Health and safety, statutory and legal management • Process management • Change management

Phase 7: production information

Is the detail 'right' for construction?

The detail of the design should be determined to enable the planning of construction including assembly and enabling works. Preferably no more changes in the design should occur after this stage. Every effort should be made to optimise the design after consideration of the whole life cycle of the product.

Before the phase

- Review membership of design teams.
- Review evaluation criteria for co-ordinated design (ideally design 100% complete).
- Review and update communications strategy.

During the phase

- Develop co-ordinated fabrication design/detail for the co-ordinated product model.
- Develop production process map for on and off-site activities for each system/work package.
- Start 'enabling works'.

Goals

- Finalise all major deliverables and proceed to the construction phase
- Gain approval to proceed through to phase 9.

Gate status

- 'Soft' gate.

Prime activity zone responsibility	Potential activity zone membership
<ul><li>Development management</li><li>Project management</li><li>Resource management</li><li>Design management</li><li>Production management</li><li>Facilities management</li><li>Health and safety, statutory and legal management</li><li>Process management</li></ul>	<ul><li>Development management</li><li>Project management</li><li>Resource management</li><li>Design management</li><li>Production management</li><li>Facilities management</li><li>Health and safety, statutory and legal management</li><li>Process management</li><li>Change management</li></ul>

Phase 8: construction

Are we ready to hand over the facility?

The design fixity and careful consideration of all constraints achieved at the previous phase should ensure the 'trouble-free' construction of the product. Any problems identified should be analysed to ensure that they do not re-occur in future projects.

Before the phase

- Finalise all major deliverables such as the project brief, business case, project execution plan, etc.
- Finalise drawings for construction along with production information.
- Ensure that all supplier bodies are in place.
- Formulate contingency plans to accommodate possible obstructive elements, e.g. weather.

During the phase

- Undertake construction works.
- Manage and monitor costs, materials, equipment and quality of suppliers' work.
- Manage the construction process.
- Review and implement hand-over plan.

- Manage health and safety.
- Liaise with stakeholders for future needs.

Goals

- Produce a building that satisfies all client requirements.
- Hand over the building as planned.

Gate status

- 'Hard' gate.

Prime activity zone responsibility	Potential activity zone membership
<ul><li>Development management</li><li>Project management</li><li>Resource management</li><li>Production management</li><li>Health and safety, statutory and legal management</li><li>Process management</li></ul>	<ul><li>Development management</li><li>Project management</li><li>Resource management</li><li>Design management</li><li>Production management</li><li>Facilities management</li><li>Health and safety, statutory and legal management</li><li>Process management</li><li>Change management</li></ul>

Phase 9: operation and maintenance

What can we learn?

The facility is handed over to the client as planned. The post-project review should identify any areas that need to be considered more carefully in future projects. The emphasis should be on creating a learning environment for everybody involved. As-built designs are documented and finalised information is deposited in the legacy archive for future use.

Before the phase

- Construct building as planned.
- Hand over the facility with all the relevant documentation.
- Store all the project information and learning lessons in the legacy archive.

- Plan for ongoing feedback from the client's organisation.
- Management team liaise with contractor team to plan handover.

During the phase

- Undertake a post-project review to examine the client's level of satisfaction.
- Examine the fulfilment of all success and performance criteria.
- Establish continuous communications with the client.
- Make ongoing review of assets with regard to:
  Functionality
  Health and Safety
  Maintaining asset information

Actions

The life cycle of the product is likely to be more than a decade. Therefore the facility life cycle should be considered and the facility examined at planned intervals either as part of the contractual arrangements or as part of continuous customer service. All lessons learned should be entered in the legacy archive and used for future projects.

Gate status

Although there are no formal gates in the process, care should be paid in establishing a programme of continuous improvement that is communicated throughout the company and the company's organisation.

Prime activity zone responsibility	Potential activity zone membership
• Development management • Project management • Resource management • Facilities management • Health and safety, statutory and legal management • Process management	• Development management • Project management • Resource management • Design management • Production management • Facilities management • Health and safety, statutory and legal management • Process management • Change management

ACTIVITY ZONES

Development management

Development management is responsible for creating and maintaining business focus which satisfies both relevant organisational and stakeholder objectives and constraints throughout the project.

For example, a proposed speculative office development needs to satisfy the developer's objectives (say, return on capital) and constraints (say, available finance) as well as fulfilling other stakeholder considerations (say, compliance with prevailing planning concerns).

The development management activity zone is likely to include the following parties:

- Senior client representation.
- Suppliers of finance to the client.
- Professional advisors.

(Obviously the supply of finance and professional advice can originate from both in-house and outside the client organisation).

Potential activity zone participation
Project management
Resource management
Design management
Production management
Facilities management
Health and safety, statutory and legal management
Process management
Change management

Project management

Project management is responsible for effectively and efficiently implementing the project to agreed performance measures in close collaboration with process management.

Performance criteria are ultimately based on requirements set out in the business case and project brief.

Project management is an agent of the development management activity zone.

Project management is ultimately responsible for preparing the project execution plan and ensuring that all relevant inputs from other activity zones are guided and integrated towards the successful implementation of the project.

The project management activity zone is likely to consist of project management professionals.

Potential activity zone participation
Development management Resource management Design management Production management Facilities management Health and safety, statutory and legal management Process management Change management

Resource management

Resource management is responsible for the planning, co-ordination, procurement and monitoring of all financial, human and material resources. (Development management establishes the overall budget.)

The resources management activity zone is likely to include the following parties:

- Quantity surveying – which will define plant and material needs and monitor their cost.
- Buying – which will procure plant and materials defined by the quantity surveying.
- Project management – which will define human resources requirements.
- Human resources – which will procure human resources defined by project management.

Potential activity zone participation
Development management Project management Design management Production management Facilities management Health and safety, statutory and legal management Process management Change management

Design management

Design management is responsible for the design process that translates the business case and project brief into an appropriate product definition. It guides and integrates all design input from other activity zones.

Potential activity zone participation
• Design professionals • Suppliers of materials/components • Main contractor and subcontractors and representatives from: 　　Production management activity zone 　　Facilities management activity zone 　　Development management activity zone 　　Project management activity zone 　　Health and safety, statutory and legal management

Production management

Production management is responsible for ensuring the optimal solution for the buildability of the design, the construction logistics and organisation for delivery of the product.

The production management activity zone is likely to include the following parties:

- Suppliers.
- Main contractor and subcontractors and representatives from:
    Design management activity zone.
    Project management activity zone.
    Health and safety, statutory and legal management.

Potential activity zone participation
Development management Project management Resource management Design management Facilities management Health and safety, statutory and legal management Process management Change management

Facilities management

Facilities management is responsible for ensuring the cost-efficient management of assets and the creation of an environment that strongly supports the primary objectives of the building owner and/or user.
The facilities management activity zone is likely to include the following parties:

- Facilities management professionals.
- Building maintenance professionals.
- Building services professionals and representatives from:
     Design management activity zone.

Potential activity zone participation
Development management
Project management
Resource management
Design management
Facilities management
Health and safety, statutory and legal management
Process management
Change management

Health and safety, statutory and legal management

Health & safety, statutory and legal management is responsible for the identification, consideration and management of all regulatory, statutory and environmental aspects of the project.

Potential activity zone participation
Development management
Project management
Resource management
Design management
Production management
Facilities management
Process management
Change management

Process management

Process management develops and operationalises the Process Protocol and is responsible for planning and monitoring each phase. Process management is an agent of the development management activity zone. Responsibilities include:

- Formulate the process execution plan, in close collaboration with project management.
- Review the phase review plan(s) and reports.
- Determine and examining the inputs and outputs of the process in terms of the deliverables at each phase.
- Offer expert recommendations to the development management activity zone with regards to the satisfactory execution of the process for delivery of the product.

The process management activity zone should consist of construction professionals who are independent of the project.

Potential activity zone participation
Development management
Project management
Resource management
Design management
Production management
Facilities management
Health and safety, statutory and legal management
Change management

Change management

Change management is responsible for effectively communicating project changes to all relevant activity zones and the development and operation of the legacy archive. Responsibilities include:

- Receiving and structuring change information.
- Distributing appropriate change information to relevant activity zones in an accurate and timely fashion.
- Retrieving and distributing appropriate legacy archive information to relevant activity zones.
- To reviewing and, where appropriate, modifying and/or updating the legacy archive.

Note: The roles of project, process and change management may be combined and this will be dependent on the size of the project.

Potential activity zone participation
Development management
Project management
Resource management
Design management
Production management
Facilities management
Health and safety, statutory and legal management
Process management

List of Abbreviations

2D	two dimensional
3D	three dimensional
AI	artificial intelligence
ATLAS	Architecture, Methodology and Tools for Computer-Integrated Large-Scale Engineering
BAA	British Airport Authority
BIM	building information model
BOT/BOOT	build-own-operate-transfer
BPF	British Property Federation
BPR	business process re-engineering
CAD	computer-aided design
CAM	computer-aided manufacturing
CE	concurrent engineering
CFT	cross-functional teams
CI	continuous improvement
CIRIA	Construction Industry Research and Information Association
COMBINE	Computer Models for the Building Industry in Europe
COMMIT	Construction Modelling and Methodologies for Intelligent Information integration
DCP	design and construction period
DFD	data flow diagrams
DFMA	design for manufacturing–assembly
EDI	electronic data interchange
EPSRC	Engineering and Physical Sciences Research Council
ER	entity relationship
FMEA	failure mode and effects analysis
FTA	fault tree analysis
HIPO	hierarchy + input-process-output
HOQ	house of quality
IAI	International Alliance for Interoperability
ICAM	(US Air Force Program for) Integrated Computer Aided Manufacturing
ICON	Integration of Construction Information Information/Integration for Construction
IDAC	Integration in Design and Construction
IDEF	integrated definition language

IDEF0	integration definition language 0 (for function modelling)
IFC	Industry Foundation Classes
IMI	Innovative Manufacturing Initiative
IS	information systems
IT	information technology
JIT	just-in-time
KBS	knowledge-based system
M&E	monitoring and evaluation
NASA	National Aeronautics and Space Administration
NPD	new product development
OLAP	on-line analytical processing
OSCON	Open Systems for Construction
PBP	pay-back period
PDCA	plan-do-check-act (also sometimes referred to as the Demming Wheel or Shewhart Cycle)
PFMEA	process failure modes effects analysis
PDS	product design specifications
PFI	Private Finance Initiative
PPP	phased program planning
QFD	quality function deployment
R&I	research and implementation
RIBA	Royal Institute of British Architects
ROI	return on investment
SCRI	Salford Centre for Research and Innovation (University of Salford, UK)
SPACE	simultaneous prototyping for an integrated construction environment
SPC	statistical process control
STEP	Standard for Exchange of Product model data
TMO	temporary multi-organisations
TQM	total quality management
VR	virtual reality
WP	work package(s)

References

Abetti, P. A. (1994) Impact of Technology on Functional Roles and Strategies. *International Journal of Technology Management*, 9(6), 529–546.

Adam, E. E. (1992) Quality Improvement as an Operations Strategy. *Industrial Management and Data Systems*, 92(4), 3–12.

Akintoye, A. (1995) Just-In-Time Application and Implementation for Building Material Management. *Construction Management and Economics*, 13(2), 105–113.

Alarcon, L. F. & Mardones, D. A. (1998) Improving the Design – Construction Interface. In *Proceedings of the 6th International Group of Lean Construction Conference*, August, Brazil.

Ali, A. (1994) Pioneering Versus Incremental Innovation: Review and Research Propositions. *Journal of Product Innovation Management*, 11, 46–61.

Allweyer, T., Babin-Ebell, T., Leinenbach, S. & Scheer, A. W. (1996) Model based Reengineering in the European Construction Industry. In *Proceedings of CIB W78 Construction on the Information Highway Conference*, Turkey, http://delphi.kstr.lth.se/w78/

Almanik, M. S. & Alitavoli, M. (1994) A Generic IDEF-0 Model of an Integrated Product Development in CIM Environment Based on Concurrent Engineering. In *Proceedings of the 9th International Conference on Applications of Artificial Intelligence in Engineering*, July 19–21, 455–462.

Alshawi, M. (1996) Awareness Disk on the Space Integrated Environment for the Construction Industry. University of Salford, Salford.

Alshawi, M. & Aouad, G. (1995) A Structured Framework for Integrating Business and Information Technology Strategies for Construction. *Civil Engineering Systems*, 12, 249–261.

Alty, J. L. (1993) Information Technology and the Construction Industry: Another Tower of Babel. In *Management of Information Technology* (eds S. K. Mathur, P. M. Betts & K. W. Tham). World Scientific, Singapore.

Amaratunga, D. & Baldry, D. (2001) The Debate about Quantitative and Qualitative Research in the Built Environment: A Question of Method or Epistemology? In *Proceedings of the 1st International Postgraduate Research Conference in the Built and Human Environment*, March, Salford, 129–148.

Ames, C. B. & Hlavacek, J. D. (1984) *Managerial Marketing for Industrial firms*. Random House Business Division, New York.

Ammerman, E., Jung, R., Katranuschkov, P. & Scherer, R. J. (1994) *Concept of an Object-Oriented Product Model for Building Design.* Technische Universtat, Dresden.

Anderson, E. J. (1994) *Management of Manufacturing, Models and Analysis.* Addison-Wesley, Wokingham.

Anderson, R E. (1993) Can Stage-Gate Systems Deliver the Goods? *Financial Executive,* Nov–Dec, 9(6), 34–38.

Aouad, G. (1997) Integration: From a Modelling Dream into an Implementation Reality. *Proceedings of CIB W55 International Support for Building Economics,* Lake District, 7–29.

Aouad, G., Kirkham, J., Brandon, P., Brown, F., Cooper, G., Ford, S., *et al.* (1993) Information Modelling in the Construction Industry – The Information Engineering Approach. *Construction Management and Economics,* 11(5), 384–397.

Aouad, G., Betts, M., Brandon, P., Brown, F., Child, T., Cooper, G., *et al.,* (1994) *ICON Final Report,* University of Salford, July.

Aouad, G., Brandon, P., Brown, F., Child, T., Cooper, G., Ford, S., *et al.* (1995) The Conceptual Modelling of Construction Management Information. *Automation in Construction,* 3, 267–282.

Aouad, G., Brandon, P., Brown, F., Cooper, G., Ford, S., Kirkham, J., *et al.* (1996) *Integration of Construction Information (ICON): Final Report – Integrated databases for the design construction and management of construction.* University of Salford, July.

Aouad, G., Alshawi, M. & Bees, S. (1997) Priority Topics for Construction IT Research. *The International Journal of Construction IT,* Winter.

Aouad, G., Hinks, J., Cooper, R., Sheath, D., Kagioglou, M. & Sexton, M. (1998) An IT Map for the Generic Design and Construction Process Protocol. *Journal of Construction Procurement,* 4(1), 132–151.

Aouad, G., Kagioglou, M. & Cooper, R. (1999) IT in Construction: A Driver or an Enabler? *Journal of Logistics and Information Management,* 12, 130–137.

Architects' Journal (1999) AJ One Hundred. *Architects' Journal,* 29 April, 41–50.

Ardhaldjian, R. & Fahner, M. (1994) Using Simulation in the Business Process Re-engineering Effort. *Industrial Engineering,* 26(27), 60–61.

Arditi, D. & Gunaydin, H. M. (1998) Factors that Affect Process Quality in the Life Cycle of Building Projects. *Journal of Construction Engineering and Management,* 124(3), 194–203.

Armstrong, S. G. & Lockley, S. R. (1995) Modelling of Generic Document Structures and the Development of an Integrated Document Data Environment. In *Product and Process Modelling in the Building Industry* (ed. R. J. Scherer). Balkema, Rotterdam.

Ashton, C. (1997) *Strategic Performance Measurement: Transforming*

Corporate Performance by Measuring and Managing the Drivers of Business Success. Business Intelligence Ltd, Wimbledon.

Assaf, S. A., Mohammed, A. K. & Muhammad, A. H. (1995) Causes of Delay in Large Building Construction Projects. *Journal of Management in Engineering*, 11(2) 45–50.

AT&T (1993) *Moving a Design into Production.* McGraw-Hill, New York.

Atkinson, A. A., Waterhouse, J. H. & Wells, R. B. (1997) A Stakeholder Approach to Strategic Performance Measurement. *Sloan Management Review*, 38(3), 25–37.

ATLAS (1992) Architecture, Methodology and Tools for Computer-Integrated Large-scale Engineering. http://www.newcastle.research.e-c.org/esp-syn/text/7280.html

Augenbroe G. (1994) An overview of the COMBINE Project. *Proceedings from the first ECPPM Conference*, Dresden. ftp://erg.ucd.ie/public/pdfiles/combine-papers/paper1.pdf

BAA Plc. (1995) *The BAA Project Process.* BAA Plc., London.

Babbie, E. (1986) *The Practice of Social Research.* Wadsworth, Belmont, CA.

Babbie, E. (1990) *Survey Research Methods.* 2nd edn. Wadsworth, Belmont, CA.

Babcock, D. L. (1991) Project Organisation, Leadership, and Control. In *Managing Engineering and Technology.* Prentice Hall, Englewood Cliffs, NJ.

Baker, B. N., Murphy, D. C. & Fisher, D. (1983) Factors Affecting Project Success. In *Project Management Handbook* (eds D. I. Cleland & W. R. King). Van Nostrand Reinhold, New York.

Baker, M. J. (1983) *Market Development: A Comprehensive Survey.* Penguin Books, Harmondsworth, England.

Balachandra, R., Brockhoff, K. K. & Pearson, A. W. (1996) R&D Project Termination Decisions: Processes, Communication, and Personnel Changes. *Journal of Product Innovation Management*, 13, 245–256.

Ball, M. (1988) *Rebuilding Construction: Economic Change in the British Construction Industry.* T. J. Press, Padstow.

Ball, M. (1996) *Rebuilding Construction.* Routledge, London.

Ballard, G. & Koskela, L. (1998) On the Agenda of Design Management Research. In *Proceedings of the 6th International Group of Lean Construction*, August, Brazil.

Banard, C. I. (1962) *The Functions of the Executive.* Harvard University Press, Cambridge, MA.

Banwell, H. (1964) *Report of the Committee on the Placing and Management of Contracts for Building and Civil Engineering Work.* HMSO, London.

Barkley, J. (1993) Applications of Concurrent Engineering. *Institute of Electrical and Electronics Engineering*, Autotestcon, GenRad Inc., IEEE Publications, 451–457.

Barlow, J. (1996) Partnering, Lean Production and the High Performance Workplace. In *Proceedings of the 4th International Group of Lean Construction*, Birmingham, UK.

Barr, V. (1990) Six Steps to Smoother Product Design. *Mechanical Engineering*, January, 48–51.

Barrie, D. S. & Paulson, B. C. (1992) *Professional Construction Management*, 3rd edn. McGraw Hill International, New York.

Bartram, P. (1994) Re-engineering Revisited. *Management Today*, July, 62–63.

Bath University (2003) Agile Construction Initiative. http://www.bath.ac.uk/Departments/Management/

Bauly, J. A. (1994) Measures of Performance. *World Class Design to Manufacture*, 1(3), 37–40.

Baxendale, A., Dulaimi, M. & Tulley, G. (1996) Simultaneous Engineering and its Implications for Procurement. In *Proceedings of the W-92 Procurement Symposium: North meets South Developing Ideas*, University of Natal, South Africa, 21–31.

Beamon, B. (1999) Measuring Supply Chain Performance. *International Journal of Operations and Production Management*, 19(3), 275–292.

Becker, H. S. (1970) *Sociological Work: Method and Substance.* Transaction Books, Chicago.

Becker, H. S. (1986) *Writing for Social Scientists: How to Start and Finish your Thesis.* University Press of Chicago, Chicago.

Beckford, J. (1998) *Quality: A Critical Introduction.* Routledge, London.

Beer, M., Ruh, R., Dawson, J. A., McCaa, B. B. & Kavanagh, M. J. (1978) A Performance Management System: Research, Design, Introduction and Evaluation. *Personnel Psychology*, 31, 505–535.

Bendell, T., Boulter, L. & Kelly, J. (1993) *Benchmarking for Competitive Advantage.* Pitman, London.

Berger, M. A. (1983) In Defence of the Case Method: a Reply to Algyris. *Academy of Management Review*, 8(2), 329–386.

Berliner, C. & Brimson, J. A. (eds) (1988) *Cost Management for Today's Advanced Manufacturing.* Harvard Business School Press, Boston.

Bettenhausen, K. N. D. & Murnighan, J. K. (1986) The Emergence of Norms in Competitive Decision-Making Groups. *Administrative Science Quarterly*, 30, 350–372.

Betts, M. (1992) *Information Technology Planning Frameworks for Computer Integrated Construction.* Construction Economics Research Unit, School of Building and Estate Management, National University of Singapore.

Betts, M. & Ofori, G. (1993) Strategies for Technology Push: Lessons for Construction Innovation. *Construction Management and Engineering*, ACSE, 454–456.

Betts, M. & Wood-Harper, T. (1994) Re-engineering Construction: A New Management Research Agenda. *Construction Management and Economics*, 12, 551–556.

Bevan, S & Thompson, M. (1991) Performance Management at the Crossroads. *Personnel Management*, November, 36–39.

Bititci, U. S. & Carrie, A. S. (1998) *Integrated Performance Measurement Systems: Structures and Relationships.* Engineering and Physical Sciences Research Council research grant, final report, Swindon.

Bititci, U. S., Carrie, A. S. & McDevitt, L. (1997) Integrated Performance Measurement Systems: An Audit and Development Guide. *The TQM Magazine*, 9(1), 46–53.

Bititci, U. S., Turner, T. & Begemann, C. (2000) Dynamics of Performance Measurement Systems. *International Journal of Operations and Production Management*, 20(6), 692–704.

BIW (2001) http:///www.biw.com

Bjork, B.C. (1989) Basic Structure of a Proposed Building Project Model. *Computer Aided Design*, 21(2), 71–78.

Boag, D. A. & Rinholm, B. L. (1989) New Product Management Practices of Small High Technology Firms. *Journal of Product Innovation Management*, 6, 109–122.

Boer, H. (1994) Flexible Manufacturing Systems. In *New Wave Manufacturing Strategies* (ed. J. Storey). Paul Chapman, London.

Bonaccorsi, A. & Lipparini, A. (1994) Strategic Partnerships in New Product Development: An Italian Case Study. *Institute of Electrical and Electronic Engineering Management Review*, Winter, 22(4), 38–46.

Bond, T. C. (1999) The Role of Performance Measurement in Continuous Improvement. *International Journal of Operations and Production Management*, 19(12), 1318–1334.

Bonsdorff, C. Von. & Andersin, H. E. (1995) Supporting the Business Process Paradigm by Means of Performance Measurements. In *Proceedings of CE96 Conference on Concurrent Engineering: A Global Perspective*, University of Vancouver, August, 151–153.

Booz, Allen & Hamilton (1982) *New Products for the 1980s*. Booz, Allen and Hamilton Inc., New York.

Bourne, M. & Nccly, A. (1998) Why do Performance Measurement Initiatives Succeed and Fail? Performance Measurement – Theory and Practice. *Papers from the 1st International Conference on Performance Measurement* (ed. A. Neely), Cambridge, July, 165–172.

Bourne, M., Mills, J., Wilcox, M., Neely, A. & Platts, K. (2000) Designing, Implementing and Updating Performance Measurement Systems. *International Journal of Operations and Production Management*, 20(7), 754–771.

Bowley, M. (1966) *The British Building Industry*. Cambridge University Press, Cambridge.

Bradley, P., Browne, J., Jackson, S. & Jagdev, H. (1995) Business Process Reengineering (BPR) – a Study of the Software Tools Currently Available. *Computers in Industry*, 25, 309–330.

Brandon, P.S. & Betts, M. (1995) *Integrated Construction Information*. E & F N Spon, London.

Brierly, H. M. (1994) The Art of Relationship Management. *Direct Marketing*, September, 22–24.

Brignal, S. (1992) *Performance Measurement Systems as Change Agents: A Case for Further Research*. Warwick Business School Research Papers, No. 72, December.

British Property Federation (1983) *Manual of the BPF System for Building Design and Construction*. British Property Federation, London.

British Standards Institute (1989) *Guide to Monitoring Product Design*. HMSO, London.

Bromley, D. B. (1986) *The Case Study Method in Psychology and Related Disciplines*. Wiley, Chichester.

Brown, A. (1996a) *Construction Modelling and Methodologies for Intelligent Information Integration*. University of Salford, Salford. http://www.salford.ac.uk/iti/projects/commit/commit.html

Brown, A. (1996b) The Architecture and Implementation of a distributed computer integrated environment. *W78'96 Published Papers*. http://delphi/kstr.lth.se/w78/,1996

Brown, M. (1996c) *Keeping Score: Using the Right Metrics to Drive World Class Performance*. Quality Resources, New York.

Brown, M. G. (1994) Is Your Measurement System Well Balanced? *Journal for Quality and Participation*, 17, 6–11.

Brown, S. L. & Eisenhardt, K. M. (1995) Product Development: Past Research, Present Findings, and Future Directions. *Academy of Management Review*, 20(2), 343–378.

Browne, J. & McMahon, C. (1993) *CADCAM, From Principles to Practice*. Addison-Wesley Publishing Company, Reading, MA.

Browne, J., Harhen, J. & Shivnan, J. (1988) *Production Management Systems: A CIM Perspective*. Addison-Wesley Publishing Company, Wokingham.

Browne, J., Sackett, P. J. & Wortmann, J. C. (1995) Future Manufacturing Systems – Towards the Extended Enterprise. *Computers in Industry*, 28(3), 235–254.

Bruce, M. & Biemens, W. G. (eds) (1995) *Product Development: Meeting the Challenge of the Design Marketing Interface*. Wiley, New York.

Bryman, A. (1993) *Quantity and Quality in Social Research*. Routledge, London.

Buffa, E. S. & Sarin, R. K. (1987) *Modern Production/ Operations Management*, 8th edn. Wiley, New York.

Building (1998) Top 200 Consultants. *Building*, 2nd October, 40–56.

Building (1999) The Improving Performance of the UK Construction Industry: Report Published for National Construction Week. *Building*, April, 1–18.

Building (2000) Joining the Dot.Coms. *Building*, 18th February, 3–4.

Building IT 2000 (1991) Construction Industry Computing Association, London.

Bulletpoint. (1996) *Creating a Change Culture – not about Structures, but Winning Hearts and Minds.* Wesley, New York.

Burbidge, J. L. (1990) Production Control: a Universal Conceptual Framework. *Production Planning and Control* 1(1), 3–16.

Burbidge, J. L. (1996) *Period Batch Control.* Oxford University Press, Oxford.

Burdett, J. O. (1994) TQM and Re-engineering: The Battle for the Organisation of Tomorrow. *The TQM Magazine*, 6(2), 7–13.

Burgess, R. (1993) *In the Field: an Introduction to Field Research.* London, Routledge.

Busby, J. S. & Williamson, A. (2000) The Appropriate Use of Performance Measurement in Non-Production Activity. *International Journal of Operations and Production Management*, 20(3), 336–358.

Butler, R. (1994) What You Measure is What You Get – An Investigation into the Measurement of the Value Added by the Purchasing Department. In *Proceedings of the 4th International Annual Conference in the Service Sector in Manufacturing Procurement*, University of Birmingham, April, 11–12.

Cameron, K. S. (1986) A Study of Organisational Effectiveness and its Predictors. *Management Science*, 32(1), 87–112.

Camp, R. C. (1989) *Benchmarking – The Search for Industry Best Practices that Lead to Superior Performance.* ASQS Quality Press, Milwaukee.

Cannon-Bowers, J. A., Oser, R. & Flanagan, D. L. (1990) Work Teams in Industry: A Selected Review and Proposed Framework. In *Teams: Their Training and Performance* (eds R. W. Swezey & E. Salas). Ablex Publishing Corporation, Norwood, NJ.

Carr, D. K & Johansson, H. J. (1995) *Best Practices in Reengineering.* McGraw-Hill, Oxford.

Carty, G. J. (1995) Construction. *Journal of Construction Engineering and Management*, September, 319–328.

Cavaye, A. L. M. (1996) Case Study Research: A Multi-Faceted Research Approach for IS. *Information Systems Journal*, 6(3), 227–242.

CBPP (1998) *Construction Industry Key Performance Indicators 1998 – Project Delivery and Company Performance.* Construction Best Practice Programme, Watford.

CBPP (1999a) *Construction Best Practice Programme.* http://www.cbpp.org.uk

CBPP (1999b) *Measure your Performance against these Key Performance Indicators: All Construction.* Construction Best Practice Programme wall chart, London.

CBPP (2002) *Respect for People: Key Performance Indicators.* Construction Best Practice Programme, Watford.

CCF (1998). http:///ww.fga-construction-procurement.co.uk/p2.html

Chadwick, B. A., Bahr, H. M. & Albrecht, S. L. (1984) The Quality Era. *Journal of Geography in Higher Education*, 8(2), 163–167.

Champy, J. (1995) *Reengineering Management: The Mandate for New Leadership.* Harper Collins, London.

Chan, P. S. & Peel, D. (1998) Causes and Impact of Re-engineering. *Business Process Management Journal*, 4(1), 45–55.

Chandler, A. D. (1997) *The Visible Hand – Managerial Revolution in American Business.* Harvard University Press, Boston, MA.

Chang, L. (1991) A Methodology for Measuring Construction Productivity. *Cost Engineering*, 3(10), 19–25.

Cherns, A. B. & Bryant, D. T. (1984) Studying the Client's Role in Construction Management. *Construction Management and Economics*, 2(6), 177–184.

Cheung, Y. (1988) Process Analysis Techniques and Tools for Business Improvements. *Business Process Management*, 4(4), 274–290.

Child, P. (1991) *The Management of Complexity.* Sloan Management Review, Autumn, 73–80.

Childe, S., Maull, R. & Mills, B. (1996) UK Experience in Business Process Re-engineering. Grant No.GR/K67328, University of Plymouth. http://bprc.warwick.ac.uk/rc-rep-9.html

Choi, C. F. & Chan, S. L. (1997) Business Process Re-engineering: Evocation, Elucidation and Exploration. *Business Process Management Journal*, 3(1), 39–63.

CIB (1996a) *Constructing a Better Image.* A report by Working Group 7: Construction Industry Board. Thomas Telford, London.

CIB (1996b) *Successful construction: Code of Practice for clients of the construction industry.* CIB, Netherlands.

CIRIA (1995) *Planning to Build?* Special Publication 113. Construction Industry Research and Information Association.

Clark, F. (1996) *Leadership for Quality: Strategies for Action.* McGraw-Hill, Maidenhead.

Clark, K. B. (1989) Project Scope and Project Performance: The Effect of Parts Strategy and Supplier Involvement on Product Development. *Management Science*, October, 35(10), 1247–1263.

Clark, K. B. (1994) Private Finance Initiatives. *New Builder & Civil Engineer*, July, 3–5.

Clark, K. B. & Fujimoto, T. (1991) *Product Development Performance,*

Strategy, Organisation and Management in the World Auto Industry. Harvard Business Press, Boston, MA.

Clark, K. B. & Wheelwright, S. C. (1993) *Managing New Product and Process Development.* The Free Press, New York.

Clark, L. A. & Zirner, U. (1993) How to Design, Develop and Implement Successful Performance Measurement Systems. *Quality and Productive Management*, 10(3), 61–80.

Coates, J. (1997) *Topical Issues – Performance Management.* CIMA Publishing, London.

Cobb, I. (1993) *JIT and the Management Accountant: A Study of Current UK Practice.* CIMA Publishing, London.

Cohen, L & Manion, L. (1994) *Research Methods in Education.* Routledge, London.

Cohodas, M J. (1988) Make the Most of Supplier Know-How. *Electronics Purchasing*, 38–39.

Colvin, H. M. (1975) *The History of the King's Work.* HMSO, 3, 1485–1660.

Compton, W. D. (1992) Benchmarking. In *Manufacturing Systems: Foundations of World-Class Practice* (eds J. A. Heim & W. D. Compton). National Academy Press, Washington DC.

Construct IT (2003) http://www.construct-it.salford.ac.uk/

Construction News (1999) Top 100 Contractors. *Construction News*, 19 April, 19–22.

Contract Journal (2000) Honesty – Waring's Best Policy. *Contract Journal*, 1 March, 14–15.

Cook, H. E. (1992) Manufacturing Systems, Foundations of World-Class Practice. In *Organising Manufacturing Enterprises for Customer Satisfaction.* National Academy of Engineering, National Academy Press, Washington, DC.

Coolican, H. (1990) *Research Methods and Statistics in Psychology.* Hodder and Stoughton, London.

Cooper, R. G. (1982) New Product Success in Industrial Firms. *Industrial Marketing Management*, 11, 215–223.

Cooper, R. G. (1983) The Impact of New Product Strategies. *Industrial Marketing Management*, 12, 243–256.

Cooper, R. G. (1984a) New Product Strategies: What Distinguishes the Top Performers? *Journal of Product Innovation Management*, 2, 151–164.

Cooper, R. G. (1984b) The Performance Impact of Product Innovation Strategies. *European Journal of Marketing*, 18(5), 223–229.

Cooper, R. G. (1988) Predevelopment Activities Determine New Product Success. *Industrial Marketing Management*, 17, 237–247.

Cooper, R. G. (1990) Stage-Gate System: A New Tool for Managing New Products. *Business Horizons*, May-June, 44–54.

Cooper, R. G. (1992) The NewProd System: The Industry Experience. *Journal of Product Innovation Management*, 9, 113–127.

Cooper, R. G. (1993) *Winning at New Products: Accelerating the Process from Idea to Launch.* Addison-Wesley, Reading, MA.

Cooper, R. G. (1994) Third-Generation New Product Processes. *Journal of Product Innovation Management*, 10, 6–14.

Cooper, R. G. (1999) From Experience: The Invisible Success Factors in Product Innovation. *Journal of Production Innovation Management*, 16, 115–33.

Cooper, R. G. & Kleinschmidt, E. J. (1986) An Investigation into the New Product Process: Steps, Deficiencies, and Impact. *Journal of Product Innovation Management*, 3, 71–85.

Cooper, R. G. & Kleinschmidt, E. J. (1987a) New products: What Separates Winners from Losers? *Product Innovation Management Journal*, 4, 169–184.

Cooper, R. G. & Kleinschmidt, E. J. (1987b) Success Factors in Product Innovation. *Industrial Marketing Management Journal*, 7, 9–21.

Cooper, R. G. & Kleinschmidt, E. J. (1987c) Success Factors in Product Innovation. *Industrial Marketing Management*, 16, 215–223.

Cooper, R. G. & Kleinschmidt, E. J. (1988) Resource Allocation in the New Product Process. *Journal of Product Innovation Management*, 17, 249–262.

Cooper, R. G. & Kleinschmidt, E. J. (1991) New Product Processes at Leading Industrial Firms. *Industrial Marketing Management*, 20, 137–147.

Cooper, R. G. & Kleinschmidt, E. J. (1994) Determinants of Timeliness in Product Development. *Journal of Product Innovation Management*, 11(5), 381–396.

Cooper, R. G. & Kleinschmidt, E. J. (1995) Benchmarking the Firm's Critical Success Factors in New Product Development. *Journal of Product Innovation Management*, 12, 374–391.

Cooper, R., Kagioglou, M., Aouad, G., Hinks, J., Sexton, M. & Sheath, D. (1998) Development of a Generic Design and Construction Process. *European Conference on Product Data Technology*, BRE, 205–214.

Coughlan, P. D. (1991) Differentiation and Integration: The Challenge of New Product Development. In *Proceedings of the 5th Annual Conference of the British Academy of Management*, 28 June .

Craig, A. & Hart, S. (1991) *Where to Now in New Product Development Research?* University of Strathclyde, Glasgow.

Crawford, C. M. (1977a) *New Products Management.* Irwin, Burr Ridge, IL.

Crawford, C. M. (1977b) Product Development: Today's Most Common Mistakes. *University of Michigan Business Review*, 6, 7–8.

Crawford, C. M. (1979) New Product Failure Rates – Facts and Fallacies. *Research Management*, Sept, 9–13.

Crawford, C. M. (1984) Protocol: New Tool for Product Innovation. *Journal of Product Innovation Management*, 1, 85–91.

Crawford, C. M. (1992) The Hidden Costs of Accelerated Product Development. *Journal of Product Innovation Management*, 9(3), 161–176.

Crawford, C. M. (1994 *New Products Management*, 4th edn. Irwin, Burr Ridge, IL.

Crawford, K. M. & Fox, J. F. (1990) Designing Performance Measurement Systems for Just-In-Time Operations. *International Journal of Production Research*, 28(11), 2025–2036.

Crittenden, V. L. (1992) Close the Marketing/ Manufacturing Gap. *Sloan Management Review*, Spring, 41–52.

Curtin University of Technology (1996) http://rolf.ece.curtin.edu.au/clive/concurrent/coneng.html

Das, T. H. (1983) Qualitative Research in Organisational Behaviour. *Journal of Management Studies*, 20(3), 311–325.

Davenport, D., Short, J. E. & Price, G. (1990) The New Industrial Engineering: Information Technology and Business Process Redesign. *Sloan Management Review*, 31(4), 11–27.

Davenport, T. H. (1993) *Process Innovation – Reengineering Work through Information Technology.* Harvard Business School Press, Boston, MA.

Davenport, T. H. (1994) Reengineering: Business Change of Mythic Proportions. *MIS Quarterly*, 6, 121–127.

Davies, A. J. & Kochhar, A. K. (1999) Why British Companies don't do Effective Benchmarking. *Integrated Manufacturing Systems*, 10(1), 26–32.

Davies-Cooper, R. & Jones, T. (1995) The Interfaces Between Design and Other Key Functions in Product Development. In *Product Development: Meeting the Challenge of the Design-Marketing Interface* (eds M. Bruce & W. G. Biemans). John Wiley & Sons Ltd, New York.

Davis, J. S. (1988) New Product Success and Failure: Three Case Studies. *Industrial Marketing Management*, 17, 103–109.

De Meyer, A. (1992) An Empirical Investigation of Manufacturing Strategies in European Industry. In *Manufacturing Strategy* (ed. C. A. Voss). Chapman & Hall, Oxford.

De Toni, A. & Tonchia, S. (1996) Lean Organisation, Management by Process and Performance Measurement. *International Journal of Operations and Production Management*, 16(2), 221–236.

De Toni, A., Nassimbeni, G. & Tonchia, S. (1995) An Instrument for Quality Performance Measurement. *International Journal for Production Economics*, 38, 199–207.

Dean, E. B. (1996) Concurrent engineering from the perspective of competitive advantage. http://www.;arc/masa/gpv

Debenham, M. (2001) Effective Process Management. *Quality World*, 7, 40–45.

Delargy, M. (1999) IT: Electronic Document Exchange – Off The Shelf. *Building*, 23 July, 58–59.

Delmar, D. (1985) *Operations and Industrial Management: Designing and Managing for Productivity.* McGraw-Hill, New York.

DeLuzio, M. C. (1993) Management Accounting in a Just-In-Time Environment. *Journal of Cost Management*, 6(4), 6–15.

Denzin, H. (1984) *The Research Act.* Prentice Hall, Englewood Cliffs.

Desa, S. & Schmitz, J. M. (1991) Development and Implementation of a Comprehensive Concurrent Engineering Method: Theory and Application. *Society of Automotive Engineers (SAE) Transactions*, Carnegie Mellon University, 100(5), 1041–1049.

Deschamps, J.-P. & Ranganath Nayak, P. (1995) *Product Juggernauts: How Companies Mobilize to Generate a Stream of Market Winners*. Harvard Business School, Boston, MA.

DeToro, I. & McCabe, T. (1997) How to stay Flexible and Elude Fads. *Quality Progress*, 30(3), 55–60.

Devinny, T. M. (1995) Significant Issues for the Future of Product Innovation. *Journal of Product Innovation Management*, 12, 70–75.

Dick, W. & Akintoye, A. (1996) Private Finance Initiative Procurement Method. In *Proceedings of the 6th COBRA Conference*, Royal Institution of Chartered Surveyors, London.

Dixon, J. R. & Arnold, P. (1994) Business Process Re-engineering: Improving in New Strategic Directions. *California Management Review*, 36(4), 93–107.

Dixon, J. R., Nanni, A. J. & Vollmann, T. E. (1990) *The New Performance Challenge – Measuring Operations for World-Class Competition.* Dow Jones-Irwin, Homewood.

Dolan, R. J. (1993) *Managing the New Product Development Process.* Addison-Wesley, Reading, MA.

Donnellon, A. (1993) Cross-Functional Teams in Product Development: the Structure to the Process. *Product Innovation Management Journal*, 18(6), 102–108.

Dorf, R. C. & Bishop, R. H. (1995) *Modern Control Systems*, 7th edn. Addison-Wesley Publishing, Chicago.

dos Santos, A. (1999) *Application of Production Management Flow Principles in Construction Sites.* Unpublished PhD thesis, University of Salford, Salford.

dos Santos, A., Powell, J. & Formoso, C. T. (1999) Evaluation of Current use of Production Management Principles in Construction Practice. In *Proceedings of the 7th International Group of Lean Construction*, University of California, Berkeley, 73–84.

Dougherty, D. (1992) Interpretive Barriers to Successful Product Innovation in Large Firms. *Organisation Science*, May, 3(2), 179–202.

Dowlatshahi, S. (1994) A Comparison of Approaches to Concurrent Engineering. *The International Journal of Advanced Manufacturing Technology*, 9(2), 106–113.

Driva, H. (1997) *The Role of Performance Measurement During Product Design and Development in a Manufacturing Environment*. PhD Thesis, Department of Manufacturing, Engineering and Operations, University of Nottingham.

DTI (1994) *Time for real Improvement: Learning from Best Practice in Japanese Construction R&D*. Department of Trade and Industry, HMSO.

DTI (1996) *Small and Medium sized Enterprise (SME) Statistics for the United Kingdom*. A Publication of the Government Statistical Service, July.

DTI (1998) *The 1998 UK R&D Scoreboard*. Department of Trade and Industry, Company Reporting, Edinburgh.

Duberley, J., Johnson, P., Cassel, C. & Close, P. (2000) Manufacturing Change: The Role of Performance Evaluation and Control Systems. *International Journal of Operations and Production Management*, 20(4), 427–440.

Dubois, A. M., Flynn, J., Verhoef, M. H. G. & Augenbroe, G. (1995) Conceptual Modelling Processes in the COMBINE Project.

Duck, J. D. (1993) Managing Change: The Art of Balancing. *Harvard Business Review*, November, 110–118.

Dwyer, L. & Mellor, R. (1991) New Product Process Activities and Project Outcomes. *R&D Management*, January, 21(1), 31–42.

Earl, M. J., Sampler, J. L. & Short, J. E. (1995) Strategies for Business Process Reengineering: Evidence from Field Studies. *Journal of Management Information Systems*, 12(1), 31–56.

Easterby-Smith, M., Thorpe, R. & Lowe, A. (1991) *Management Research Introduction*. Sage, London.

Eccles, R. G. & Pyburn, P. J. (1992) Creating a Comprehensive System to Measure Performance. *Management Accounting*, October, 41–44.

Edwards, C. & Peppard, J. W. (1994) Business Process Redesign: Hype, Hope or Hypocrisy? *Journal of Information Technology*, 9, 251–266.

Egan, J. (1998) *Rethinking Construction*. Report from the Construction Task Force, Department of the Environment, Transport and Regions, UK.

Eisenhardt, K. M. (1989) Building Theories from Case Study Research. *Academy of Management Review*, 14(41), 532–550.

Elmuti, D. & Kathawala, Y. (1997) An Overview of Benchmarking Process: A Tool for Continuous Improvement and Competitive Advantage. *Benchmarking for Quality Management and Technology*, 4(4), 229–243.

Elzinga, D. J., Horak, T., Chung-Yee, L. & Bruner, C. (1995) Business Process Management: Survey and Methodology. *IEEE Transactions on Engineering Management*, 24(2), 119–128.

Emmerson, H. (1962) *Studies of Problems before the Construction Industries*. HMSO, London.

EPSRC (1998) http://www.epsrc.ac.uk

Ettlie, J. E. (1990) Methods that Work for Integrating Design and Manufacturing. In *Managing the Design-Manufacturing Process* (eds J. E. Ettlie & H. W. Stoll). McGraw-Hill, New York.

Ettlie, J. E. & Stoll, H. W. (1990) *Managing the Design-Manufacturing Process*. McGraw-Hill, New York.

Eureka, W. E. (1987) Introduction to Quality Function Deployment. In *Quality Function Deployment: A Collection of Presentations and QFD Studies*, American Suppliers Institute, Dearborn, MI, January.

Evangelidis, K. (1992) Performance Measured is Performance Gained. *The Treasurer*, February, 45–47.

Evans, J. R. (1993) *Applied Productions and Operations Management*, 4th edn. West Publishing, St Paul, MN.

Evans, J. R. & Laskin, R. L. (1994) The Relationship Marketing Process: A Conceptualisation and Application Approach. *Industrial Marketing Management*, 23(5), 439–452.

Fact Finders (1996) *Southwestern Bell Telephone Product Development Benchmarking Study*. Fact Finders Incorporated, California.

Fairclough, J. (2002) *Rethinking Construction Innovation and Research: A Review of Government R&D Policies and Practices*. Department for Transport, Local Government and the Regions, HMSO.

Fazoop, P., Moselhi, O., Theberge, P. & Revay, S. (1988) Design Impact of Construction Fast-Track. *Construction Management and Economics*, 6(3), 195–208.

Fellows, R. & Lui, A. (1997) *Research Methods for Construction*. Blackwell Science, Oxford.

Fenves, S.J., Flemming, U., Hendrickson, C. Maher, M. L. & Schmitt, G. (1990) Integrated Software Environment for Building Design and Construction. *Computer-Aided Design*, 22(1), 27–36..

Fiedler, K. D., Grover, V. & Teng, J. T. C. (1994) Information Technology-Enabled Change: The Risks and Rewards of Business Process Redesign and Automation. *Journal of Information Technology*, 9, 267–275.

Filippini, R., Forza, C. & Vinelli, A. (1998) Trade-Off and Compatibility Between Performance: Definitions and Empirical Evidence. *International Journal of Production Research*, 36(12), 3379–3406.

Fitzgerald, L. & Moon, P. (1996) *Performance Measurement in Service Industries: Making it Work*. The Chartered Institute of Management Accountants, London.

Fitzgerald, L., Johnston, R., Brignall, S., Silvestro, R. & Voss, C. (1991) *Performance Measurement in Service Industries: Making it Work.* The Chartered Institute of Management Accountants, London.

Flapper, S. D. P., Fortuin L. & Stoop, P. P. M. (1996) Towards Consistent Performance Management Systems. *International Journal of Operations and Production Management,* 16(7), 27–37.

Fleet, T. (1995) Partnering in the Construction Industry 1 – Contractual Issues. *Construction Law,* 6(5), 175–177.

Fleury, A. (1995) Quality and Productivity in the Competitive Strategies of Brazilian Industrial Enterprises. *World Development,* 23(1), 73–85.

Flynn, B. B., Schroeder, R. D. & Sakakibara, S. (1994) A Framework for Quality Management Research and an Associated Measurement Instrument. *Journal of Operations Management,* 11, 339–366.

Fogarty, D. W., Hoffmann, T. R. & Stonebraker, P. W. (1989) *Production and Operations Management.* South Western Publishing, Ohio.

Ford Automotive Company (1988) *FMEA manual, Design FMEA, Process FMEA.* Brentwood.

Formoso, C. T. (1997) Improvement Example from the 1990s. In *Proceedings of the International Transfer of Construction Management Practices Workshop,* June, Oslo, Norway.

Fowler, A. (1990) Performance Management: The MBO of the 90s? *Personnel Management,* July, 47–51.

Franks, J. (1990) *Building Procurement Systems: a Guide to Building Project Management,* 2nd edn. CIOB, Ascot.

Froese, T. & Paulson, B. (1994) OPIS: An Object Model-Based Project Information System. *Microcomputers in Civil Engineering,* 9, 113–28.

Fry, T. D. (1995) Japanese Manufacturing Performance Criteria. *International Journal of Production Research,* 33(3), 933–954.

Fry, T. D. & Cox, J. F. (1989) Manufacturing Performance: Local versus Global Measures. *Production and Inventory Management Journal,* second quarter, 52–57.

Fryer, B. (1997) *The Practice of Construction Management,* 3rd edn. Blackwell Science, Oxford.

Gaafar, H. K. & Perry, J. G. (1998) Limitation of Design Liability for Contractors. *International Journal of Project Management,* 17(5), 301–308.

Gable, G. (1994) Integrating Case Study and Survey Research Methods: An Example in Information Systems. *European Journal of Information Systems,* 3(2), 112–126.

Gadd, K. W. (1995) Business Self-Assessment: A Strategic Tool for Building Process Robustness and Achieving Integrated Management. *Business Process Re-engineering and Management Journal,* 1(3), 66–85.

Galbraith, J. R. & Kazanjian, R. K. (1986) *Strategy Implementation: Structure, Systems and Process,* 2nd edn. West, St Paul.

Galhenage, G. P. (1994) *Just in time manufacturing.* Department of Computer Engineering, Curtin University of Technology. http://rolf.ece.curtin.edu.au/clive/jit/jit.html

Galhenage, G. P. (1996) *Comparison of traditional engineering and CE approach.* Department of Computer Engineering, Curtin University of Technology.

Gallimore, P., Williams, W. & Woodward, D. (1997) Perceptions of Risk in the Private Finance Initiative. *Journal of Property Finance,* 8(2), 164–176.

Galsworth, G. D. (1997) *Visual Systems: Harnessing the Power of a Visual Workplace.* AMACOM, Bristol.

Ghalayini, A. M. & Noble, J. S. (1996) The Changing Basis of Performance Measurement. *International Journal of Operations and Production Management,* 16(8), 63–80.

Gilgeous, V. & Gilgeous, M. (1999) A Framework for Manufacturing Excellence. *Integrated Manufacturing Systems,* 10(1), 33–44.

Gills, J. & Johnson, P. (1997) *Research Methods for Managers,* 2nd edn. Paul Chapman Publishing, London.

Globerson, S. (1985a) Issues in Developing a Performance Criteria System for an Organisation. *International Journal of Production Research,* 23(4), 639–646.

Globerson, S. (1985b) *Performance Criteria and Incentive Systems.* Elsevier, Bristol.

Goldense, B. L. (1994) Rapid Product Development. *World Class Design to Manufacture,* 1(1), 21–28.

Goold, M. (1991) Strategic Control in the Decentralised Firm. *Sloan Management Review,* winter, 69–81.

Goold, M. & Quinn, J. J. (1990) The Paradox of Strategic Controls. *Strategic Management Journal,* 11, 43–57.

Gorbett, J. (1986) Design for Economic Manufacture. *Annals of CIRP,* 35(1), 93.

Green, F. B., Amenkhienan, F. & Johnson, G. (1991) Performance Measures and JIT. *Management Accounting,* 72(8), 50–53.

Gregory, M. J. (1993) Integrated Performance Measurement: A Review of Current Practice and Emerging Trends. *International Journal of Production Economics,* 30, 281–296.

Griffin, A. (1992) Evaluating QFD's Use in U.S. Firms as a Process for Developing Products. *Journal of Product Innovation Management,* September, 9(3), 171–187.

Griffin, A. (1993) Metrics for Measuring Product Development Cycle Time. *Journal of Product Innovation Management,* 10, 112–125.

Griffin, A. (1997) PDMA Research on New Product Development Practices: Updating Trends and Benchmarking Best Practices. *Journal of Product Innovation Management,* 14, 429–458.

Griffin, A. & Hauser, J. R. (1996) Integrating R&D and Marketing: A Review and Analysis of the Literature. *Journal of Product Innovation Management*, 13, 191–215.

Group EFO (1995) *Innovation Survey: Report on New Products*. Group EFO Limited, Western.

Gunasekaran, A. & Love, P. E. D. (1998) Concurrent Engineering: A Multi-Disciplinary Approach for Construction. *Logistics Information Management*, MCB University Press, 11(5), 295–300.

Gunasekaran, A., Forker, L. & Kobu, B. (2000) Improving Operations Performance in a Small Company: A Case Study. *International Journal of Operations and Production Management*, 20(3), 359–369.

Gupta, A. K. & Wilemon, D. L. (1990) Accelerating the Development of Technology-Based New Products. *California Management Review*, 32(2), 22–32.

Gupta, A. K., Raj, S. P. & Wilemon, D. (1985) R&D and Marketing Dialogue in High-Tech Firms. *Industrial Marketing Management*, 14, 289–300.

Gyles, R. (1992) *Royal Commission into Productivity in the New South Wales Building Industry*. Government Printer, London.

Hackman, J. R. & Wageman, R. (1995) Total Quality Management: Empirical, Conceptual and Practical Issues. *Administrative Science Quarterly*, 40(2), 309–342.

Hague, P. (1994) *Questionnaire Design*. Kogan Page, London.

Hammer, M. & Stanton, S. A. (1995) *The Re-engineering Revolution*. Harper Business, New York.

Hammer, M. & Champy, J. (1997) *Reengineering the Corporation: A Manifesto for Business Revolution*, 5th edn. Nicholas Brealey, London.

Handfield, R. B. (1994) Effects of Concurrent Engineering on Make-To-Order Products. *IEEE Transactions on Engineering Management*, 41(4), 384–393.

Hanna, V. & Burns, N. D. (1997) *The Behavioural Implications of Performance Measures*. Macmillan Press, Houston.

Hannus, M., Karstila, K. & Tarandi, V. (1995) Requirements on Standardised Building Product Data Models and Computer Assisted Construction Planning in Total Project Systems. *International Journal of Construction Information Technology*, 5.

Harkins, J. R. & Dubreuil, M. P. (1993) Concurrent Engineering in Product Design/ Development. *Plastics Engineering*, August, 27–31.

Harper, J. (1984) *Measuring Business Performance: A Manager's Guide*. Institute of Manpower Studies, Gower, Hartlepool.

Harrington, H. J. (1991) *Business Process Improvement*. McGraw Hill, New York.

Harrington, H. J. (1999) Performance Improvement: A Total Poor-Quality Cost System. *The TQM Magazine*, 11(4), 221–230.

Hart, S. (1996) New Product Success: Measurement, Methodologies, Models and Myths. *Hidden Versus Open Rules in Product Development: DELFT*, April, 151–168.

Hartley, J. & Mortimer, J. (1990) *Simultaneous Engineering: The Management Guide.* Industrial Newsletters Ltd, Dunstable.

Harvey, J. P. (1971) *The Master Builders – Architecture in the Middles Ages.* Thames & Hudson, London.

Harvey-Jones, J. (1991) *Making it Happen: Reflections on Leadership.* Collins, London.

Hay, E. (1990) Implementing JIT Purchasing: Phase III – Selection. *Production and Inventory Management Review* with *APICS News*, 10(3), 28–29.

Hayes, R. H. & Abernathy, W. J. (1980) Managing our way to Economic Decline. *Harvard Business Review*, July-August, 67–77.

Hayes, R. H., Wheelwright, S. C. & Clark, K. B. (1988) *Dynamic Manufacturing: Creating the Learning Organisation.* Free Press, New York.

Haynes, I. & Frost, N. (1994) Accelerated Product Development: An Experience with Small and Medium-sized Companies. *World Class Design to Manufacture*, 1(5), 32–37.

Hendry, L. C. (1998) Applying World Class Manufacturing to Make-to-Order Companies: Problems and Solutions. *International Journal of Operations and Production Management*, 18(11), 1086–1100.

Herbsman, Z. & Ellis, R. (1991) Research of Factors Influencing Construction Productivity. *Construction Management and Economics*, 8(1), 49–61.

Hibberd, P & Djebarni, R. (1996) Criteria of Choice for Procurement Methods. *Proceedings of COBRA 96*, University of the West of England.

Higgins, G. & Jessop, N. (1993) *Communications in the Construction Industry.* National Joint Consultative Committee of Architects Quantity Surveyors and Builders, Tavistock Institute, London.

Hill, T. J. (1991) *Production and Operations Management: Text and Cases*, 2nd edn. Prentice Hall, London.

Hill, T. J. (1992) Incorporating Manufacturing Perspectives in Corporate Strategy. In *Manufacturing Strategy* (ed. C. A. Voss). Chapman & Hall, Oxford.

Hinks, J., Aouad, G., Cooper, R., Sheath, D., Kagioglou, M. & Sexton, M. (1997) IT and the Design and Construction Process: A Conceptual Model of Co-maturation. *International Journal of Construction*, July, 56–62.

Hise, R. T., O'Neal, L., Parasuraman, A. & McNeal, J. U. (1990) Marketing/R&D Interaction in New Product Development: Implications for New Product Success Rates. *Journal of Product Innovation Management*, 7(2), 142–155.

HM Treasury (1998) *Innovating for the Future.* Department of Trade and Industry, HMSO, London.

Hoffman, L, R. (1979) *The Group Problem Solving Process: Studies of a Valance Model.* Praeger, New York.

Hofstede, G. H. (1968) *The Game of Budget Control.* Tavistock, Cambridge.

Holmes, G. (1994) Putting Concurrent Engineering into Practice. *World Class Design to Manufacture*, 1(5), 38–42.

Hopkins, D. (1980) New Products Winners and Losers.*The Conference Board*, 4–9, New York.

Hopp, W. J., Spearman, M. L. & Woodruff, D. L. (1990) Practical Strategies for Lead-Time Reduction. *Manufacturing Review*, 3(2), 78–84.

Hopwood, A. G. (1984) *Accounting and Human Behaviour.* Prentice-Hall, Englewood Cliffs.

Horna, J. (1994) *The Study of Leisure.* Oxford University Press, Oxford.

Horne, C. A. (1987) Product Strategy and the Competitive Advantage. *P&IM Review*, December.

House, C. H. & Price, R. L. (1991) The Return Map: Tracking Product Teams. *Harvard Business Review*, January-February, 92–100.

Howard, H. C. (1992) *Linking Design Data with Knowledge-Based Construction Systems.* CIFE Spring Symposium, March, 1–24.

Howell, D. (1999) Builders get the Manufacturers In. *Professional Engineer*, May, 24–25.

Hughes, G. D & Chafin, D. C. (1996) Turning New Product Development into a Continuous Learning Process. *Journal of Product Innovation Management*, 13, 89–104.

Hughes, W. (1991) Modelling the construction process using plans of work. Construction Project Modelling and Productivity – *Proceedings of an International Conference CIB W65*, Dubrovnik.

Hunt, J. (1995) Who's Sold on the PFI. *Property Week*, May, 20–21.

Huthwaite, B. (1988) Designing in Quality. *Quality*, 27(11), 34–35.

IDEF (2002) IDEF Family of Methods http://www.idef.com

Imai, M. K. (1986) *The Key to Japan's Competitive Success.* McGraw-Hill, New York.

Imai, K., Nonaka, I. & Takeuchi, H. (1985) Managing the New Product Development Process: How Japanese Companies Learn and Unlearn. In *The Uneasy Alliance; Managing the Productivity-Technology Dilemma* (1989) (eds K. B. Clark, R. H. Hayes & C. Lorenz). Harvard Business School Press, Boston, MA.

Inwood, D. & Hammond, J. (1993) *Product Development: An Integrated Approach.* Kogan Page Ltd, NY.

IT2005 (1995) *Building on IT2005.* CICA, Cambridge.

Jaafari, A. (1997) Concurrent Construction and Life Cycle Project Management. *Journal of Construction Engineering and Management*, 123(4), 427–436.

Jackson, J. & Hall, D. (1992) Speeding Up: New Product Development. *Management Accounting*, October, 265–268.

Jacobs, L. & Herbig, P. (1998) Japanese Product Development Strategies. *Journal of Business & Industrial Marketing*, 13(2), 132–154.

Jamieson, I. A. (1997) Development of a Construction Process Protocol to Promote a Concurrent Engineering Environment within the Irish Construction Industry. *First International Conference on Concurrent Engineering in Construction.* Institution of Structural Engineers Informal Study Group on Computing in Structural Engineering, 4–5 July.

Jassawalla, A. R. & Sashittal, H. C. (1998) An Examination of Collaboration in High-Technology New Product Development Processes. *Journal of Product Innovation Management*, 15, 237–254.

Jick, T. D. (1979) Mixing Qualitative and Quantitative Methods: Triangulation Action. *Administrative Science Quarterly*, 11, 339–366.

Johansson, H. J., McHugh, P., Pendlebury, A. J. & Wheeler, W. A. (1993) *Business Process Re-engineering: Breakpoint Strategies for Market Dominance.* John Wiley & Sons, Chichester.

Johne, A. & Snelson, P. (1996) *Successful Product Development Management Practices in American and British Firms.* Basil Blackwell, Massachusetts.

Johnson, H. T. (1994) Relevance Regained: Total Quality Management and the Role of Management Accounting. *Critical Perspectives on Accounting*, 5(2), 259–267.

Johnson, H. T. & Kaplan, R. S. (1987) *Relevance Lost – The Rise and Fall of Management Accounting.* Harvard Business School, Boston.

Jones, D. T. (1992) Beyond the Toyota Production System: The Era of Lean Production. In *Manufacturing Strategy: Process and Content* (ed. C. Voss) pp.188–210. Chapman & Hall, New York.

Jones, I. (1978) *Mixing Qualitative Research and Quantitative Methods on Sports Fan Research.* http://www.nova.edu/ssss/QR/QR3-4/nau.html

Juran, J. M. (1988) *Juran on Planning for Quality*, 2nd edn. Free Press, New York.

Juran, J. M. (1989) *Juran on Leadership for Quality: An Executive Handbook.* Free Press, New York.

Juran, J. M. (1992) *Juran on Quality by Design.* Free Press, New York.

Kadefors, A. (1999) Client – Contractor Relations: How Fairness Considerations and Interests Influence Contractor Variation Negotiations. In *Proceedings of the 7th International Group of Lean Construction*, 231–240. University of California, Berkeley.

Kagioglou, M. (1999) *Adapting Manufacturing Project Processes into Construction: A Methodology.* Unpublished PhD thesis, University of Salford, Salford.

Kagioglou, M., Cooper, R., Aouad, G., Hinks, J., Sexton, M. & Sheath, D.

(1998a) *Final Report: Generic Design and Construction Process Protocol.* University of Salford, Salford.

Kagioglou, M., Cooper, R., Aouad, G., Hinks, J., Sexton, M. & Sheath, D. (1998b) *A Generic Guide to the Design and Construction Process Protocol.* University of Salford.

Kagioglou, M., Cooper, R., Aouad, G., Hinks, J., Sexton, M. & Sheath, D. (1998c) Cross-Industry Learning: The Development of a Generic Design and Construction Process Based on the Stage/Gate New Product Development Process found in the Manufacturing Industry. In *Proceedings of the Engineering Design Conference*, Brunel, Uxbridge.

Kaplan, R. S. (1984) Yesterday's Accounting Undermines Production. *Harvard Business Review*, 62, 95–101.

Kaplan, R. S. & Norton, D. P. (1992) The Balanced Scorecard: Measures that Drive Performance. *Harvard Business Review*, January-February, 71–79.

Kaplan, R. S. & Norton, D. P. (1993) Putting the Balanced Scorecard to Work. *Harvard Business Review*, September-October, 134–147.

Kaplan, R. S. & Norton, D. P. (1996) *The Balanced Scorecard: Translating Strategy into Action.* Harvard Business School Press, Boston.

Karagozoglu, N. & Brown, W. B. (1993) Time-Based Management of the New Product Development Process. *Journal of Product Innovation Management*, 10, 204–215.

Kartam, N. (1994) ISICAD: Interactive System for Integrating CAD and Computer-Based Construction Systems. *Microcomputers in Civil Engineering*, 9, 41–51.

Kartam, N. A. (1996) Making Effective Use of Construction Lessons Learned in Project Life Cycle. *Journal of Construction Engineering and Management*, March, 14–21.

Katzenbach, J. (1996) *Real Change Leaders.* Nicholas Brealey, London.

Kaydos, W. (1991) *Measuring, Managing and Maximising Performance.* Productivity Press, Portland OR.

Keegan, D. P., Eiler, R. G. & Jones, C. R. (1989) Are Your Performance Measures Obsolete? *Management Accounting*, June, 45–50.

Keppel, G. (1991) *Design and Analysis: a Researcher's Handbook*, 3rd edn. Prentice Hall, Englewood Cliffs, NJ.

Kerssens-van Drongelen, I. C. (1998) Different Performance Measurement Procedures for Different Purposes. In *Performance Measurement – Theory and Practice*. (ed. A. Neely). Papers from the 1st International Conference on Performance Measurement, Cambridge, July, 584–591.

Khurana, A. & Rosenthal, S. R. (1998) Towards Holistic 'Front Ends' in New Product Development. *Journal of Product Innovation and Management*, 15, 57–74.

Kim, B. O. (1994) Business Process Re-engineering: Building a Cross-

Functional Information Architecture. *Journal of Systems Management*, 45(12), 30–35.

King, N. (1996) *The Qualitative Research Review*. Sage, London.

Kirk, J. (2001) Business Improvement: The Continuous Cycle. *Quality World*, 7, 37–39.

Klapper, L. S., Hamblin, N., Hutchinson, L., Novak, L. & Vivar, J. (1999) *Supply Chain Management: A Recommended Performance Measurement Scorecard*. Logistics Management Institute, McLean, VA.

Klein, M. M. (1994a) Reengineering Methodologies and Tools. *Information System Management*, Spring, 30–35.

Klein, M. M. (1994b) The Most Fatal Reengineering Mistakes. *Information Strategy: The Executive's Journal*, 10(4), 21–28.

Kmetovicz, R. E. (1992) *New Product Development: Design and Analysis*. Wiley-Interscience, New York.

Knutt, E. (2000) Learn to Share. *Building*, 7 April, 54–55.

Kohler, N. & Bedell, J.R. (1995) Building Project Model for Life Cycle Applications. In *Product and Process Modeling in the Building Industry* (ed. R. J. Scherer). Balkema, Rotterdam.

Kornelius, L. & Wamelink, J. W. F. (1998) The Virtual Corporation: Learning from Construction. *Supply Chain Management*, 3(4), 193–202.

Koskela, L. (1991) State of the Art of Construction Robotics in Finland. In *Proceedings of the 8th International Symposium on Automation and Robotics in Construction*, Stuttgart, 65–70.

Koskela, L. (1992) *Application of the New Production Philosophy to Construction*. Technical report no. 72. Centre for Integrated Facility Engineering, Stanford University, CA.

Koskela, L. (1999) Management of Production in Construction: A Theoretical View. In *Proceedings of the 7th International Group of Lean Construction*, pp. 241–252. University of California, Berkeley.

Koskela, L. & Houvila, P. (1996) *On Foundations of Concurrent Engineering*. VTT Building Technology, Espoo, Finland.

Kosekla, L., Melles, B. & Wamelink, J. W. F. (1992) *Theme Report on Production Philosophies in Other Industries*. CIB W82 report, Technical Research Centre of Finland/Delft University of Technology.

KPI Working Group (2000) *KPI Report for the Minister for Construction*. January, Department of the Environment, Transport and the Regions, The Stationery Office.

KPMG & CICA (1993) *Building on IT for Quality*. KPMG, London.

Kuczmarski, T. D. (1992) *Managing New Products: the Power of Innovation*. Prentice Hall, Englewood Cliffs, NJ.

Kuczmarski, T. D. (1994) *Winning New Product and Service Practices for the 1990s*. Kuczmarski & Associates, Chicago.

Kumaraswamy, M. M. & Chan, D. W. M. (1998) Contributors to Construction Delays. *Construction Management and Economics Journal,* 16(1), 17–29.

Kuwaiti, M. E. & Kay, J. M. (2000) The Role of Performance Measurement in Business Process Re-engineering. *International Journal of Operations and Production Management,* 20(12), 1411–1426.

Lahdenperra, P. (1995) *Reorganising the Building Process: the Holistic Approach.* VTT Building Technology, Espoo, Finland.

LaPlante, A. & Alter, A.E. (1994) Corning Inc.: The Stage-gate Innovation Process. Computerworld, October 31, 24 (44), 81.

Latham, M. (1994) *Constructing the Team: Joint Review of Procurement and Contractual Arrangements in the UK Construction Industry.* Department of the Environment, HMSO.

Lee, A., Cooper, R. & Aouad, G. (2000) Production Management: The Process Protocol Approach. *Journal of Construction Procurement,* 6(2), 164–183.

Lee, A., Marshall-Ponting, A. J., Aouad, G. Wu, S., Koh, W. W. I., Fu, C. *et al.* (2003) *Developing a Vision of nD-enabled Construction.* Construction IT, University of Salford, Salford.

Lee, R. G. & Dale, B. G. (1998) Business Process Management: A Review and Evaluation. *Business Process Management Journal,* 4(3), 214–225.

Letza, S. R. (1996) The Design and Implementation of the Balanced Scorecard: An Analysis of Three Companies in Practice. *Business Process Re-engineering and Management Journal,* 2(3), 54–76.

Lewin, K. (1938) *The Conceptual Representation and the Measurement of Psychological Forces.* Duke University Press, Durham.

Li, H. & Love, P. E. D. (1998) Developing a Theory of Construction Problem Solving. *Construction Management and Economics,* 16, 721–727.

Lincoln, Y. S. & Guba, E. G. (1985) *Naturalistic Inquiry.* Sage, Beverley Hills, CA.

Liner, M. (1992) First Experiences Using QFD in New Product Development. *Design For Manufacture,* American Society of Mechanical Engineers, 51, 57–63.

Lingle, J. H. & Schiemann, W. A. (1996) From Balanced Scorecard to Strategy Gauge: Is Measurement Worth it? *Management Review,* March, 56–52.

Little, A. D. (1991) *The Arthur D. Little Survey on the Product Innovation Process.* Arthur D. Little, Cambridge.

Lockamy, A. (1998) Quality Focused Performance Measurement Systems: A Normative Model. *International Journal of Operations and Production Management,* 18(8), 740–766.

Lockamy, A. & Cox, J. F. (1995) An Empirical Study of Division and Plant Performance Measurement Systems in Selected World Class Manufacturing Firms: Linkages for Competitive Advantage. *International Journal of Production Research,* 33(1), 221–236.

Locke, L. F., Spirduso, W. W. & Silverman, S. J. (1993) *Proposals that Work: A Guide for Planning Dissertations and Grant Proposals.* Sage, Houston.

Lorsch, J W, and Lawrence, P R. (1965). Organizing for Product Innovation. *Harvard Business Review*, January-February, pp. 109–120.

Love, P. E. D. & Gunasekaran, A. (1996) Towards Concurrency and Integration in the Construction Industry. *International Conference on Concurrent Engineering*, Toronto, August, 26–29.

Love, P. E. D. & Li, H. (1998) From BPR to CPR – Conceptualising Reengineering in Construction. *Business Process Management Journal*, 4(4), 291–305.

Love, P. E. D. & Holt, G. D. (2000) Construction Business Performance Measurement: The SPM Alternative. *Business Process Management Journal*, 6(5), 408–416.

Love, P. E. D., Gunasekaran, A. & Li, H. (1998) Concurrent Engineering: A Strategy for Procuring Construction Projects. *International Journal of Project Management*, 6(6), 177–185.

Love, P. E. D., Smith, J. & Li, H. (1999) The Propagation of Rework Benchmark Metrics for Construction. *International Journal of Quality and Reliability Management*, 16(7), 638–658.

Low, S. P. (1993) The Application of Quantitative Methods for Just-In-Time Construction. *Building and Estate Management*, 22, 37–51.

Luck, R. & Newcombe, R. (1996) The Case for the Integration of the Project Participants' Activities within a Construction Project Environment. In *The Organisation and Management of Construction: Shaping Theory and Practice: Volume II* (eds D. A. Langford & A. Retik). Oxford University Press, Oxford.

Luck, R., McGeorge, D. & Betts, M. (1997) *Research Futures: Academic Responses to Industry Challenges*, Construct IT Centre of Excellence, Salford.

Luiten, B., Froese, T. M., Bjork, B-C, Cooper, G., Junge, R. Karstila, K., *et al.* (1993) An Information Reference Model for Architecture, Engineering and Construction. *Proceedings of the First International Conference on the Management of Information Technology for Construction* (eds K. Mathur, M. Betts & K. Tham). World Scientific and Global Publication Services, Singapore.

Lutz, J.D. & Hijabi, A. (1993) Planning Repetitive Construction: Current Practice. *Construction Management and Economics*, 11, 99–100.

Lynch, R. L. & Cross, K. F. (1991a) *Measure Up – The Essential Guide to Measuring Business Performance.* Mandarin, London.

Lynch, R. L. & Cross, K. F. (1991b) *Measure Up! Yard Sticks for Continuous Improvement.* Blackwell Business, Norwich.

Macpherson, I. (1995) Cutting Costs: Better by Half. *Building*, 10 February.

Makey, P. (1995) *Business Process Reengineering Strategies, Methods and Tools.* Butler Group, Bath.

Male, S. (1996) Comments at 'A New Way of Working', CIC Conference, Scandic Hotel, February.

Manganelli, R. L. & Klein, M. M. (1994) Your Re-engineering Toolkit. *American Management Association*, 83(8), 26–30.

Mapes, J., New, C. & Szwejczewski, M. (1997) Performance Trade-Offs in Manufacturing Plants. *International Journal of Operations and Production Management*, 17(10), 1020–1033.

Marquis, D. C. (1972) The Anatomy of Successful Innovations. *Managing Advancing Technology*, 1, 35–48. American Management Association, New York.

Marsh, R. (1996) Innovation in Small and Medium Sized Enterprises, 1995 Survey. *Economic Trends*, October, 516.

Marshall, C. & Rossman, G. B. (1989) *Designing Qualitative Research.* Sage, Houston.

Maskell, B. H. (1991) Performance Measurement for World Class Manufacturing. *Management Accounting*, May, 32–33.

Masterman, J. W. E. (1992) *An Introduction to Building Procurement Systems.* E&FN Spon, London.

Matters, M. & Evans, A. (1997) *The Nuts and Bolts of Benchmarking.* http://www.ozemail.com.au/benchmark/nuts.bolts.html

Maylor, H. (1997) Concurrent New Product Development: An Empirical Assessment. *International Journal of Operations and Production Management*, 17(12), 1196–1214.

Maylor, H. & Grosling, R. (1998) The Reality of Concurrent New Product Development. *Integrated Manufacturing Systems*, 9(2), 69–76.

McGarth, M. E. (1996) *Setting the Pace in Product Development.* Butterworth-Heinemann, Boston, MA.

McKinnon, S. M. & Bruns, W. J., Jr (1992) *The Information Mosaic.* Harvard Business School Press, Boston, MA.

McMahon, C. & Browne, J. (1993) *Just-In-Time in CAD – CAM from Principles to Practice.* Addison-Wesley, Wokingham.

McNair, C. J. & Mosconi, W. (1987) Measuring Performance in Advanced Manufacturing Environment. *Management Accounting*, July, 28–31.

McNair, C. J. & Liebfried, K. H. J. (1992) *Benchmarking: A Tool for Continuous Improvement.* Oliver Wright, Essex.

Medori, D. & Steeple, D. (2000) A framework for Auditing and Enhancing Performance Measurement Systems. *International Journal of Operations and Production Management*, 20(5), 520–533.

Meister, D. (1985) *Behaviour Analysis and Measurement Methods.* John Wiley & Sons, Cambridge.

Mello, S. & Vermette, D. (1995) Developing Breakthrough Products: How the Best in Class do it. *Management Roundtable Conference on Product Development Best Practices for Defining Customer Needs*, New Orleans.

Meyer, C. (1994) How the Right Measures Help Teams Excel. *Harvard Business Review*, May-June, 95–103.

Meyer, M. W. & Gupta, V. (1994) The Performance Paradox. In *Research in Organisational Behaviour* (eds B. M. Staw & L. L. Cummins). JAI Press, Greenwich.

Milburn, I. (1992) A Race Against Time. *Manufacturing Breakthrough*, Jan/Feb, 15–19.

Miles, M. B. (1979) Qualitative Data as an Attractive Nuisance: The Problem of Analysis. *Administrative Science Quarterly*, 24, 590–601.

Miles, M. B. & Huberman, A. M. (1984) *Analysing Qualitative Data: A Source Book for New Methods*. Sage, Houston.

Miles, R. & Ballard, G. (1997) Contracting for Lean Performance: Contracts and the Lean Construction Team. In *Proceedings of the 5th International Group of Lean Construction*, Gold Coast, Australia, July.

Mintzberg, H. (1973) *The Nature of Managerial Work*. Harper & Row, New York.

Mitchell Madison Group (1995) *New Product Development: Organisation and Process in Service Companies*. Mitchell Madison Group, New York.

Miyatake, Y. & Kangari, R. (1993) Experiencing Computer Integrated Construction. *Journal of Construction, Engineering and Management*, 119(2).

MOB (1994) Rapport Final, Modeles Objet Batiment, Appel d'Offres du Plan Construction et Architecture. *Programme Communication/Construction*, 11.

Moenaert, R. K. & Souder, W. E. (1990) An Information Transfer Model for Integrating Marketing and R&D Personnel in New Product Development Projects. *Journal of Product Innovation Management*, 7, 91–107.

Moenaert, R. K., Souder, W. E., DeMeyer, A. & Seschoolmeester, D. (1994) R&D – Marketing Integration Mechanisms, Communication Flows, and Innovativeness. *Journal of Product Innovation Management*, January, Vol. 11, No. 1, pp. 31–45.

Mohamed, S. (1996) Benchmarking and Improving Construction Productivity. *Benchmarking for Quality Management and Technology*, 3(3), 50–58.

Mohamed, S. & Yates, G. (1995) Re-engineering Approach to Construction: A Case Study. In *Proceedings of the 5th East Asia Pacific Conference on Structural Engineering and Construction*, Griffith University, Brisbane.

Mohsini, R A. (1996) Strategic Design: Front End Incubation of Buildings. *Proceeding of the 1996 Annual Conference of CIB Working Commission 92*, University of Natal, South Africa.

Mohsini, R. A. & Davidson, C. H. (1992) Detriments of Performance in the Traditional Building Process. *Journal of Construction Management and Economics*, 10, 343–359.

Mokhtar, A. & Bedard, C. (1995) Towards Integrated Construction Technical Documents: A New Approach Through Product Modeling. In *Product and Process Modeling in The Building Industry*. (ed. R. J. Scherer). Balkema, Rotterdam.

Mondon, Y. (1983) *Toyota Production System.* Industrial Engineering and Management Press, Norcross, GA.

Moore, J. (1992) Concurrent Engineering: The Quest for Quality. *Manufacturing Engineer*, February, 16–20.

Moran, J. W. & Brightman, B. K. (1998) Effective Management of Healthcare Change. *The TQM Magazine*, 10(1), 27–29.

Morenc, R. & Rangan, R. (1992) Information Management to Support Concurrent Engineering Environments. In *Computers in Engineering* (ed. G. A. Gabriel). American Society of Mechanical Engineering, 135–147.

Morse, J. (1991) Approaches to Qualitative – Quantitative Methodological Triangulation. *Nursing Research*, 40(1), 23–27.

Motwani, J. (2001) Critical Factors and Performance Measures of TQM. *The TQM Magazine*, 13(4), 292–300.

Motwani, J., Kumar, A. & Novakoski, M. (1995) Measuring Construction Productivity: A Practical Approach. *Work Study*, 44(8), 18–20.

Muckler, F. A. & Seven, S. A. (1992) Selecting Performance Measures: Objective Versus Subjective Measurement. *Human Factors*, 34, 441–455.

Muhlemann, A., Oakland, J. & Lockyer, K. (1992) *Production and Operations Management*, 6th edn. Pitman, London.

Nadler, D. A. (1980) *Feedback and Organisational Development: Using Data-Based Methods.* Addison-Wesley Publishing, Boston, MA.

Nadler, D. A. & Tushman, M. L. (1997) *Competing by Design: the Power of Organisational Architecture.* Oxford University Press, New York.

Nanni, A. J., Dixon, J. R. & Vollmann, T. E. (1990) Strategic Control and Performance Measurement – Balancing Financial and Non-Financial Measures of Performance. *Journal of Cost Management*, 4(2), 33–42.

Nau, D. (1994) *Mixing Methodologies: Can Bimodal Research be a Viable Post-Positivist Tool?* http://nova.edu/ssss/QR/QR2-3/nau.html

Neely, A. (1998) *Measuring Business Performance: Why, What and How.* Profile Books, Cranfield.

Neely, A. (1999) The Performance Measurement Revolution: Why Now and What Next? *International Journal of Operations and Production Management*, 19(2), 205–228.

Neely, A. & Adams, C. (2000) *Perspectives on Performance: The Performance Prism.* http://www.cranfield.ac.uk/som/cbp/prismarticle.pdf

Neely, A., Gregory, M. & Platts, K. (1995) Performance Measurement System Design. *International Journal of Operations and Production Management*, 15(4), 80–116.

Neely, A., Gregory, M. & Platts, K. (1996a) Performance Measurement System Design. *International Journal of Operations and Production Management*, 15(4), 80–116.

Neely, A., Mills, J., Gregory, M., Richards, H., Platts, K. & Bourne, M. (1996b) *Getting the Measure of your Business*. Findlay Publications, Holton Kirby.

Neely, A., Richards, H., Mills, J. & Platts, K. (1996c) What Makes a Good Performance Measure? In *Proceedings of the 2nd European Operations Management Association Conference*, Twente, Holland, May, 362–371.

Neely, A., Richards, H., Mills, J., Platts, K. & Bourne, M. (1997) Designing Performance Measures: A Structured Approach. *International Journal of Operations and Production Management*, 17(11), 1131–1152.

Neely, A., Mills, J., Platts, K., Richards, H., Gregory, M., Bourne, M. *et al.* (2000) Performance Measurement System Design: Developing and Testing a Process-Based Approach. *International Journal of Operations and Production Management*, 20(10), 1119–1145.

Neo, R. B. (1995) Accounting for Waste in Construction. In *Proceedings of the 1st International Conference on Construction Project Management*, Nanyang Technological University, Singapore, January, 399–406.

Nesan, L. J. & Holt, G. D. (1999) *Empowerment in Construction Organisations: The Way Forward for Performance Improvement*. Research Studies Press, Somerset.

Nevins, J. L. & Whitney, D. E. (1989) *Concurrent Design of Products and Processes: A Strategy for the Next Generation in Manufacturing*. McGraw-Hill, New York.

O'Brien, M. J. (1997) Integration at the Limit: Construction Systems. *International Journal of Construction Information Technology*, Summer, 5, 1.

O'Connor, P. (1986) New Product Ventures to Spell Success for Industrial Firms. *Marketing News*, July 4, 20(5).

O'Connor, P. (1994) FROM EXPERIENCE: Implementing a Stage Gate Process: A Multicompany Perspective. *Journal of New Product Management*, 11, 183–200.

Oakland, J. S. (1995) *Total Quality Management: The Route to Improving Performance*, 2nd edn. Butterworth Heinemann, Boston, MA.

Oh, C. J. & Park, C. S. (1993) An Economic Evaluation Model for Product Design Decisions under Concurrent Engineering, *The Engineering Economist*, Summer, 38(4), 275–297.

Olshavsky, R. W. & Spreng, R. A. (1996) An Exploratory Study of the Innovation Evaluation Process. *Journal of Product Innovation Management*, 13, 512–529.

Olson, E. M., Walker, O. C., Jr & Ruekert, R. W. (1995) Organising for Effective New Product Development: The Moderating Role of Product Innovativeness. *Journal of Marketing Research*, 59, 31–45.

O'Reilly, J. J. N. (1987) *Better briefing means better buildings*. Building Research Establishment report BR 95; B.R.E., Garston.

OSCON (1997) *Open Systems for Construction*. http://www.salford.ac.uk/iti/att/oscon.html

Otley, D. (1987) *Accounting Control and Organisational Behaviour*. Heinemann, Strassburg.

Ould, M A (1995). *Business Processes – Modelling and Analysis for Reengineering and Improvement*, Wiley & Sons, New York.

Ovenden, T. R. (1994) Business Process Re-engineering: Definitely Worth Considering. *The TQM Magazine*, 6(3), 56–61.

Page, A. L. (1994) Results from PDMA's Best Practices Study: The Best Practices of High Impact New Product Programs. In *Proceedings of the EEI/PDMA Conference on New Product Innovation*.

Pannesi, R. T. (1994) Interfacing New Product Introduction with MRP. *American Production and Inventory Control Society*, APICS, 16–17.

Paper, D. (1997) The Value of Creativity in Business Process Re-engineering. *Business Process Management*, 3(3), 218–231.

Parasuraman, A., Zeithmal, V. A. & Berry, L. L. (1990) *Delivering Service Quality: Balancing Customer Perceptions and Expectations*. Free Press, New York.

Parker, A. (1997) Engineering is not Enough. *Manufacturing Engineer*, 76(6), 267–271.

Parry, M. & Song, X. M. (1993) Determinants of R&D-Marketing Interface in High-tech Japanese Firms. *Journal of Product Innovation Management*, 10, 4–22.

Peer, S. (1986) An Improved Systematic Activity Sampling Technique for Work-Study. *Construction Management and Economics*, 4(2), 151–159.

Peltz, D. C. & Andrews, F. M. (1966) *Scientists in Organisations*, revised edn. University of Michigan Press, Ann Arbor.

Peppard, J. & Rowland, P. (1995) *The Essence of Business Process Re-engineering*. Prentice Hall, Reading.

Peters, T. J. & Watcrman, R. H. (1982) *In Search of Excellence: Lessons from American Best-Run Companies*. Harper & Row, Cambridge.

Pheng, L. S. & Tan, S. K. L. (1998) How 'Just-In-Time' Wastages can be Quantified: Case Study of a Private Condominium Project. *Construction Management and Economics*, 16, 621–635.

Phillips, T. (1950) *Report of a Working Party to the Minister of Works*. HMSO, London.

Pigford, D. & Baur, G. (1995) *Expert Systems for Business: Concepts and Applications*, 2nd edn. Boyd & Fraser, Danvers, MA.

PISTEP (1994) Process Engineering Activity Model.

Pittiglio, C., Rabin, R., Todd, P. & McGrath, M. (1995) *Product Development Leadership for Technology-Based Companies: Measurement and Management – A Prelude to Action*. Pittiglio, Rabin, Todd & McGrath, Western.

Plossl, G. W. (1987) *Engineering for the Control of Manufacturing*. Prentice-Hall, Englewood Cliffs, NJ.

Plossl, G. W. (1991) *Managing in the New World of Manufacturing*. Prentice-Hall, Englewood Cliffs, NJ.

Poirier, C. C. & Tokarz, S. J. (1996) *Avoiding the Pitfalls of Total Quality*. ASQC Quality Press, Milwaukee.

Porter, L. J., Gadd, K. W. & Oakland, J.S. (1995) Adding up to the Right Result. *Self-Assessment – The Magazine of Continuous Quality Improvement*, July, 41–45.

Powell, G. C. (1980) *An Economic History of the British Building Industry 1815-1979*. The Architectural Press, London.

Powell, J. (1995) Virtual Reality and Rapid Prototyping for Engineering. *Proceedings of the Information Technology Awareness Workshop*, January 1995, University of Salford, Salford.

Pritchard, R. D., Jones, S. D., Roth, P. L., Stuebing, K. K. & Ekeberg, S. E. (1988) Effects of Group, Goal Setting and Incentives on Organisational Productivity. *Journal of Applied Psychology*, 73, 337–358.

Pritchard, R. D., Roth, P. L., Jones, S. D. & Roth, P. G. (1991) Implementing Feedback Systems to Enhance Productivity: A Practical Guide. *National Productivity Review*, Winter, 57–67.

Process Protocol (2002) http://www.processprotocol.com

Puddicombe, M. S. (1997) Designers and Contractors: Impediments to Integration. *Journal of Construction Engineering and Management*, 123, (3), 245–252.

Pugh, S. (1991) *Total Design, Integrated Methods for Successful Product Engineering*. Addison-Wesley Publishing Company Inc., Boston, MA.

Putnam, A O. (1985) A Redesign for Engineering. *Harvard Business Review*, 63(9), 139–144.

Ramabadron, E., Dean, J. W., Jr & Evans, J. R. (1997) Benchmarking and Project Management: A Review and Organisational Model. *Benchmarking for Quality Management and Technology*, 4(1), 47–5.

Reinertsen, D. G. (1997) *Managing the Design Factory: A Product Developer's Toolkit*. The Free Press, New York.

Remenyi, D. & Williams, B. (1996) The Nature of Research: Qualitative or Quantitative, Narrow Paradigmatic. *International Journal of Information Systems*, 6(2), 131–146.

Remenyimi, D., Williams, B., Money, A. & Swartz, E. (1998) *Doing Research in Business and Management*. Sage Publications, London.

Revenaugh, D. L. (1994) Business Process Re-engineering: The Unavoidable Challenge. *Management Decision*, 32(7), 16–27.

Rezgui, Y., Brown, A., Cooper, G., Aouad, G., Kirkham, J. & Brandon, P. (1996) An Integrated Framework for Evolving Construction Models. *International Journal of Construction IT*, 4(1), 47–60.

RIBA (1980) *Handbook of Architectural Practice and Management*. RIBA Publications, London.

RIBA (1991) *Architect's Handbook of Practice Management*, 4th edn. RIBA Publications, London.

RIBA. (1997) *RIBA Plan of Work for the Design Team Operation*, 4th edn. RIBA Publications, London.

Riedel, J. C. K. H. & Pawar, K. S. (1997) The Consideration of Production Aspects During Product Design Stages. *Integrated Manufacturing Systems*, 8(4), 208–214.

Ritchie, B., Marshall, D. & Eardley, A. (1998) *Information Systems in Business*. International Thomson Business Press, New Jersey.

Robbins, S. R. (1996) *Organisational Behaviour: Concepts, Controversies and Applications*. Prentice Hall International, Houston.

Roberts, E. B. (1987) *Managing Technological Innovations: A Search for Generalizations. In Managing Technological Innovation* (ed. E. B. Roberts). Oxford University Press, Oxford.

Rochford, L. & Rudelius, W. (1992) How Involving More Functional Areas Within a Firm Affects the New Product Process. *Journal of Product Innovation Management*, 9(4), 287–299.

Rockart, J. F. (1979) Chief Executives Define Their Own Data Needs. *Harvard Business Review*, 57, March-April, 81–93.

Rolstadas, A. (1998) Enterprise Performance Measurement. *International Journal of Operations and Production Management*, 18(9), 989–999.

Rook, J. & Medhat, S. (1996) Using Metrics to Monitor Concurrent Product Development. *Industrial Management and Data Systems*, 1, 3–7.

Rosenau, M. D., Jr (1988) Phased Approach Speeds Up New Product Development. *Research & Development*, November, 52–55.

Rosenau, M. D., Jr (1990) *Faster New Product Development: Getting the Right Product to Market Quickly*. AMACOM, New York.

Rosenau, M. D., Jr, Griffin, A., Castellion, G. A. & Anschuetz, N. F. (1996) *The PDMA Handbook of New Product Development*. John Wiley and Sons Inc., New York.

Rosenthal, S. R. & Tatikonda, W. (1992) *Integrating Design and Manufacturing for Competitive Advantage: Competitive Advantage Through Design Tools and Practices*, Oxford University Press Inc., New York.

Ross, E. M. (1994) The Twenty-first Century Enterprise, Agile Manufacturing and Something Called CALS. *World Class Design to Manufacture*, 1(3), 5–10.

Rossman, G. B. & Wilson, B. L. (1991) Numbers and Words Revisited. *Evaluation Review*, 9(5), 627–643.

Rothwell, R. (1972) Factors for Success in Industrial Innovations. In *Project. SAPPHO – A Comparative Study of Success and Failure in Industrial Innovation*, Science Policy Research Unit, University of Sussex, Brighton.

Rothwell, R., Freeman, C., Horsley, A., Jervis, V. T. P., Robertson, A. & Townsend, J. (1974) SAPPHO Updated – Project SAPPHO Phase II. *Research Policy*, 3, 258–291.

Rummler, G. & Brache, A. (1990) *Improving Performance: How to Manage the White Space on the Organisational Chart*. Jossey-Bass, San Francisco.

Ruskin, C. (1995) *The Product Champion Tests his Vision*. MCB University Press, Cambridge.

Russell, R. (1992) The Role of Performance Measurement in Manufacturing Excellence. In *Proceedings of the 27th Conference on Annual British Production and Inventory Control*, November, Birmingham.

Ryan, H. W. (1994) Reinventing the Business. *Information Systems Management*, 11(2), 77–79.

Salomone, T. A. (1995) *What Every Engineer Should Know About Concurrent Engineering*. Marcel Dekker, Inc., New York.

Sanchez, A. M. & Elola, L. N. (1991) Product Innovation Management in Spain. *Journal of Product Innovation Management*, March, 8(1), 49–56.

Sanger, M. (1998) Supporting the Balanced Scorecard. *Work Study*, 47(6), 197–200.

Sanvido, V. E. (1988) Conceptualisation of Construction Process Model. *Journal of Construction Engineering and Management*, 114, 294–310.

Sanvido, V. E. (1990) *An integrated building process model*. Technical report No 1 Computer integrated construction research program. Dept of Architectural engineering, Pennsylvania State University.

Sapoutzis, P. (1995) *Use of Modern Manufacturing Techniques to Improve the Operation of a Production Cell*. Unpublished MSc Advanced Manufacturing Systems dissertation, University of Salford, Salford.

Sarin, S. & Kapur, G. M. (1990) Lessons from Product Failures: Five Case Studies. *Industrial Marketing Management*, 19, 301–313.

Sayer, A. (1984) *Method in Social Sciences*. Routledge, New York.

Schmenner, R. W. & Vollmann, T. E. (1994) Performance Measures: Gaps, False Alarms and the 'Usual Suspects.' *International Journal of Operations and Production Management*, 14(12), 58–69.

Schmidt, J. B. (1995) New Product Myopia. *Journal of Business & Industrial Marketing*, Winter, 10(1), 23–34.

Schonberger, R. J. (1982) *Japanese Manufacturing Techniques: Nine Hidden Lessons in Simplicity*. Free Press, New York.

Schonberger, R. J. (1990) *Building a Chain of Customers.* Free Press, New York.

Schonberger, R. J. (1996) *World Class Manufacturing: The Next Decade.* Free Press, New York.

Schoonhoven, C. B., Eisenhardt, K. M. & Lyman, K. (1990) Speeding Products to Market: Waiting Time to First Product Introduction in New Firms. *Administrative Science Quarterly*, 35, 177–207.

Schriver, W. & Bowlby, R. (1984) Changes in Productivity and Composition of Output in Building Construction. *Review of Economics and Statistics*, 67, 318–322.

Scott, G. M. (1995) Downsizing Business Process Re-engineering and Quality Improvement Plans: How are they Related? *Information Strategy: The Executive Journal*, 11(3), 18–34.

Senge, P. M. (1990) *The Fifth Discipline: The Art and Practice of the Learning Organisation.* Century Business, London.

Serpell, A. (1999) Integrating Quality Systems in Construction Projects: The Chilean Case. *International Journal of Project Management*, 17(5), 317–322.

Serpell, A. & Alarcon, L. F. (1998) Construction Process Improvement Methodology for Construction Projects. *International Journal of Project Management*, 16(4), 215–221.

Sethi, A. K. & Sethi, S. P. (1990) Flexibility in Manufacturing: A Survey. *International Journal of Flexible Manufacturing Systems*, 2(4), 289–328.

Seymour, D. & Rooke, J. (1995) The Culture of the Industry and the Culture of Research. *Construction Management and Economics*, 13(6), 511–523.

Shadur, M. A., Rodwell, J. J. & Bamber, G. J. (1995) Factors Predicting Employees' Approval of Lean Production. *Human Relations*, 48(12), 1403–1423.

Shahat, A. M. & Rosowsky, D. V. (1995) Accounting for Human Error During Design and Construction. *Journal of Architectural Engineering ASCE*, 121(2), 84–92.

Sheath , D. M., Woolley, H., Cooper, R., Hinks, J. & Aouad, G. (1996) A Process for Change: The Development of a Generic Design and Construction Process Protocol for the UK Construction Industry. In *Proceedings of the CIT Conference*, Institute of Civil Engineers, April, Sydney, Australia.

Shingo, S. (1989) *A Study of the Toyota Production System from an Industrial point of View* (trans. A. P. Dillon). Productivity Press, Japan.

Shinohara, I. (1988) *New Production System: JIT Crossing Industry Boundaries.* Productivity Press, Cambridge, MA.

Sidwell, A. C. (1990) Project Management: Dynamics and Performance. *Construction Management and Economics*, 8, 159–178.

Simister, S. (1995) Case Study Methodology for Construction Management Research. In *Proceedings of the 11th Annual ARCOM Conference*, York, 4, 16–18.

Simmonds, K. (1981) Strategic Management Accounting. *Management Accounting*, April, 26–29.

Simon, E. (1944) *The Placing and Management of Building Contracts*. HMSO, London.

Simons, R. (1995) Control in the Age of Empowerment. *Harvard Business Review*, 73(2), 80–88.

Sinclair, D. & Zairi, M. (1995a) Effective Process Management through Performance Measurement: Part II – Benchmarking Total Quality Based Performance Measurement for Best Practice. *Business Process Re-engineering and Management Journal*, 1(2), 58–72.

Sinclair, D. & Zairi, M. (1995b) Effective Process Management through Performance Measurement: Part III – An Integrated Model of Total Quality Based Performance Measurement. *Business Process Re-engineering and Management Journal*, 1(3), 50–65.

Sink, D. S. (1985) *Productivity Management: Planning, Measurement and Evaluation, Control and Improvement*. John Wiley & Sons, Milton Keynes.

Sink, D. S. (1986) Performance and Productivity Measurement: The Art of Developing Creative Scoreboards. *Industrial Engineer*, January, 86–90.

Sink, D. S. & Tuttle, T. C. (1989) *Planning and Measurement in your Organisation of the Future*. Industrial Engineering and Management Press, Norcross, GA.

Skibniewski, M. & Molinski, J. (1989) Modeling of Building Production Activities for Multi-Facility Projects. *Construction Management and Economics*, 7 357–365.

Skinner, W. (1974) The Decline, Fall, and Renewal of Manufacturing. *Industrial Engineering*, October, 32–38.

Skoyles, E. R. (1976) *Material Wastage – A Misuse of Resources*. Building Research Establishment, Garston.

Skoyles, E. R. (1978) *Site Accounting for Materials*. Building Research Establishment, Garston.

Slack, N. (1991) *The Manufacturing Advantage*. Mercury Books, London.

Slade, B N. (1993) *Compressing the Product Development Cycle*. American Management Association, New York.

Smith, P. G. & Reinertsen, D. G. (1991) *Developing Products in Half the Time*. Van Nostrand Reinhold, New York.

Sommerville, J. & Stocks, B. (1996) Realising the Client's Strategic Requirements: Motivating Teams. *Proceedings of COBRA '96*, University of the West of England, Bristol.

Song, X. M. & Parry, M. E. (1997) The Determinants of Japanese New Product Success. *Journal of Marketing Research*, 34, 64–76.

Song. X. M., Thieme, R. J. & Xie, J. (1998) The Impact of Cross-Functional Joint Involvement across Product Development Stages: An Exploratory Study. *Journal of Product Innovation Management*, 15, 289–303.

Souder, W. E. (1987) *Managing New Product Innovations*. Lexington Books, Lexington, MA.

Southampton University (2003) http://dsse.ecs.soton.ac.uk/

Sower, V. E., Motwani, J. & Savoie, M. J. (1997) Classics in Production and Operations Management. *International Journal of Operations and Production Management*, 17(1), 15–28.

Spekman, R. (1988) Strategic Supplier Selection: Understanding Long-term Buyer Relationships. *Business Horizons*, 31(4), 75–81.

Spivey, W. A, Munson, J. M. & Wolcott, J. H. (1997) Improving the New Product Development Process: a Fractal Paradigm for High-Technology Products. *Journal of Product Innovation Management*, 14, 203–218.

Stake, R. (1995) *The Art of Case Research*. Sage Publications, Thousand Oaks, CA.

Stalk, G. E. P. & Shulman, L. (1992) Competing on Capabilities: The New Rules of Corporate Strategy. *Harvard Business Review*, March/ April, 57–69.

Stauffer, R. N. (1988) Converting Customers to Partners at Ingersoll. *Manufacturing Engineering*, September, 41–44.

Steier, F. (1991) Introduction: Research as Self-Reflexivity – Self-Reflexivity as a Social Process. In *Research and Reflexivity* (ed. F. Steier), 1–11. Sage Publications, Hemel Hempstead.

Stoll, H. W. (1986) Design for Manufacture: An Overview. *Applied Mechanical Review*, 39(9), 1356–1364.

Strauss, A. & Corbin, J. (1990) *Basics of Qualitative Research: Grounded Theory Procedures and Techniques*. Sage Publications, Newbury Park.

Sumanth, D. J. (1985) *Productivity Engineering and Management*. McGraw-Hill, Singapore.

Susman, G, I. (1992) *Integrating Design and Manufacture for Competitive Advantage*. Oxford University Press, New York.

Susman, G. I. & Dean, J. W., Jr (1992) Development of a Model for Predicting Design for Manufacturability Effectiveness. In *Integrating Design and Manufacturing for Competitive Advantage* (ed. G. Susman). Oxford University Press, New York.

Swenson, D. W. & Cassidy, J. (1993) The Effect of JIT on Management Accounting. *Journal of Cost Management for the Manufacturing Industry*, 7(1), 39–47.

Swink, M. L., Sandvig, C. & Mabert, V. A. (1996) Customizing Concurrent Engineering Processes: Five Case Studies. *Journal of Product Innovation Management*, 13, 229–244.

Syrett, M. & Lammiman, J. (1997) *From Lean to Fitness: Developing Corporate Muscle*. Cromwell Press, Trowbridge.

Taguchi, G. & Wu, Y. (1980) *Introduction to Off-Line Quality Control.* Central Japan Quality Association, Nagoya.

Takeuchi, H & Nonaka, I. (1986) *The New Product Development Game.* Harvard Business Press, Boston, MA.

Takeuchi, H. & Quelch, J. A. (1983) Quality is more than making a Good Product. *Harvard Business Review*, 61(4), 139–145.

Taleb-Bendiab, A. (1993). The Concurrent Engineering Approach. *Lecture Notes*, Manchester Metropolitan University, Manchester.

Teicholz, P & Fischer, M. (1994) Strategy for Computer Integrated Construction Technology. *Journal of Construction, Engineering and Management*, 120(1).

Tellis, W. (1997a) Introduction to the Case Study. *The Qualitative Report*, 3(2). http://www.nova.edu/ssss/qr/qr3-2/tellis1.html

Tellis, W. (1997b) Application of a Case Study Methodology. *The Qualitative Report*, 3(3). http://www.nova.edu/ssss/qr/qr3-3/tellis2.html

Tellis, W. (1997c) Results of a Case Study on Information Technology at a University. *The Qualitative Report*, 3(4). http://www.nova.edu/ssss/qr/qr3-4/tellis3.html

Teng, J. T. C., Grover, V. & Fiedler, K. D. (1994) Business Process Reengineering: Charting a Strategic Path for the Information Age. *California Management Review*, 37(7), 9–31.

Thomas, R. J. (1993) *New Product Development: Managing and Forecasting for Strategic Success.* The Portable MBA Series. John Wiley and Sons, New York.

Thomas, R. J. (1995) *New Product Success Stories: Lessons from Leading Innovators.* John Wiley & Sons, New York.

Tidd, J., Bessant, J. & Pavitt, K. (1997) *Managing Innovation.* John Wiley & Sons, Chichester.

Toni, A. de & Tonchia, S. (2001) Performance Measurement Systems. *International Journal of Operations and Production Management*, 21(1/2), 46–70.

Trafford Park Development Corporation Manchester (1997) *Company Census of the Trafford Park Urban Development Area.* Clucas Ward Forster, Manchester.

Trygg, L. (1993) Concurrent Engineering Practices in Selected Swedish Companies: A Movement or an Activity of the Few? *Journal of Product Innovation Management*, 10, 403–415.

Tsang, A. H. C., Jardine, A. K. S. & Kolodny, H. (1999) Measuring Maintenance Performance: A Holistic Approach. *International Journal of Operations and Production Management*, 19(7), 691–715.

Tucker, R. L., O'Connor, J.T., Gatton, T. M., Gibson, G.E., Haas, C. T. & Hudson, D. N. (1994) *The Impact of Computer Technology on Construction's Future.* Department of Civil Engineering, University of Texas at Austin, Texas.

Turino, J. (1990) From Design For Test to Concurrent Engineering. *Institute of Electrical and Electronic Engineering*, 345–349.

Turney, P. B. B. & Anderson, B. (1989) Accounting for Continuous Improvement. *Sloan Management Review*, 30(2), 37–48.

Tutesigensi, A. (1999) *Managing Project Identification: An Investigation into the Effective Management of the Identification Phase of Building Projects in Uganda.* Unpublished PhD thesis, University of Leeds, Leeds.

Ulrich, K. T. & Eppinger, S. D. (1995) *Product Design and Development.* McGraw-Hill, Singapore.

United Nations (1959) *Government Policies and the Cost of Building.* ECE, Geneva.

Upton, D. (1998) Just-In-Time and Performance Measurement Systems. *International Journal of Operations and Production Management*, 18(11), 1101–1110.

Utterback, J. M. (1971) The Process of Technological Innovation Within the Firm. *Academy of Management Journal*, March, 75–88.

Vakola, M. & Rezgui, Y. (2000) Critique of Existing Business Process Re-engineering Methodologies: The Development and Implementation of a New Methodology. *Business Process Management Journal*, 6(3), 238–250.

Venegas, P. & Alarcon, L. F. (1997) Selecting Long-Term Strategies for Construction Firms. *Journal of Construction Engineering and Management*, 123(4), 388–398.

Venetucci, R. (1992) Benchmarking: A Reality Check for Strategy and Performance Objectives. *Production and Inventory Management Journal*, 33(4), 32–36.

Villemeur, A. (1992). *Reliability, Availability, Maintainability and Safety Assessment, Methods and Techniques*, Vol. 1. John Wiley & Sons, New York.,

Vokurka, R. J., Choobineh, J. & Vadi, L. (1996) A Prototype Expert System for the Evaluation and Selection of Potential Suppliers. *International Journal of Operations and Production Management*, 16(12), 106–127.

Vonderembse, M. A. & White, G. P. (1996) *Operations Management: Concepts, Methods and Strategies.* West Publishing, New York.

Voordijk, J. (1994) *Towards Integrated Logistics in Supply Chains: Developments in Construction.* University Press, Maastricht.

Voss, C. A. (ed.) (1995) *Manufacturing Strategy: Process and Content.* Chapman & Hall, London.

Voss, C. A., Ahlstrom, P. & Blackmon, K. (1997) Benchmarking and Operational Performance: Some Empirical Results. *International Journal of Operations and Production Management*, 17(10), 1046–1058.

Voss, C. A., Russell, V. & Twigg, D. (1990) Implementation Issues in Simultaneous Engineering. *International Journal of Technology Management*, 6(4), 293–302.

VTT Finland (2003) http://www.vtt.fi/rte/

Wallen, N. E. & Frankel, J. R. (1991) *Educational Research: a Guide to the Process.* McGraw Hill, New York.

Walsh, W. J. (1990) Get the Whole Organisation behind New Product Development. *Research Technology Management*, 33(6), 32–36.

Ward, A. C., Liker, J. K., Cristiano, J. J. & Sobek, I. I. D. K. (1994) Set-Based Concurrent Engineering and Toyota. *American Society of Mechanical Engineering: Design Theory and Methodology*, DTM'94, DE, 68, 79–90.

Ward, A. C., Liker, J. K., Cristiano, J. J. & Sobek, I. I. D. K. (1995) The Second Toyota Paradox: How Delaying Decisions Can Make Better Cars Faster. *Sloan Management Review*, Spring, 43–61.

Ward, S. C., Curtis, B. & Chapman, C. B. (1991) Objectives and Performance in Construction Projects. *Journal of Construction Management and Economics*, 9, 343–353.

Waterman, R. H. (1987) *The Renewal Factor.* Bantam Press, Chicago.

Waters, M. (1995) *Globalisation.* Routledge, London.

Wesner, J. W., Hiatt, J. M. & Trimble, D. C. (1995) *Winning With Quality: Applying Quality Principles in Product Development.* Addison-Wesley Publishing Company.

Wheelwright, S. C. & Clark, K. B. (1992) *Revolutionising Product Development: Quantum Leaps in Speed, Efficiency and Quality.* Free Press, New York.

White, A. (1996) *Continuous Quality Improvement: A hands-on Guide to setting up and sustaining a Cost Effective Quality Programme.* Judy Piatkus, Gloucester.

Whitelaw, J, (1996) World Ambitions: Leading Edge Procurement and Project Management take BAA into the 21st Century. *New Civil Engineer*, Supplement 5/1996, 18–23.

Winch, G. & Campagnac, E. (1995) The Organisation of Building Projects: An Anglo/ French Comparison. *Construction Management and Economics*, 13, 3–14.

Winch, G., Usmani, A. & Edkins, A. (1998) Towards Total Project Quality: A Gap Analysis Approach. *Construction Management and Economics*, 16, 193–207.

Wisner, J. D. & Fawcett, S. E. (1991) Link Firm Strategy to Operating Decisions through Performance Measurement. *Production and Inventory Management Journal*, third quarter, 5–11.

Wnuk, A. J. (1990) Some Topics on a Model for the Design Process. Institute of Electrical and Electronic Engineering. *Proceedings of AI Simulation and Planning in High Autonomy Systems*, 26–27 March, 149.

Womack, J. P. & Jones, D. T. (1996a) *Beyond Toyota: How to Root Out Waste and Pursue Perfection.* Harvard Business Press, Boston, MA.

Womack, J. P. & Jones, D. T. (1996b) *Lean Thinking.* Simon & Schuster, Oxford.

Womack, J. P., Jones, D. T. & Roos, D. (1990) *The Machine that Changed the World.* Rawson Associates, New York.

Yan, H. S. & Jiang, J. (1999) Agile Concurrent Engineering. *Integrated Manufacturing Systems*, 10(2), 103–112.

Yin, K. (1984) *Case Study Research.* Sage Publications, Beverly Hills.

Yin, R. (1989) *Case Study Research.* Sage, Newbury.

Young, S. M. & Selto, F. H. (1991) New Manufacturing Practices and Cost Management: A Review of the Literature and Directions for Research. *Journal of Accounting Literature*, 10, 265–298.

Yung, W. K. C. (1997) A Stepped Composite Methodology to Redesign Manufacturing Processes through Re-engineering and Benchmarking. *International Journal of Operations and Production Management*, 17(4), 375–388.

Zairi, M. (1991) *Total Quality Management for Engineers.* Gulf Publishing Company, Houston.

Zairi, M. (1994) *Measuring Performance for Business Results.* Chapman & Hall, London.

Zairi, M. (1995) The Integration of Benchmarking and BPR: A Matter of Choice or Necessity? *Business Process Re-engineering and Management Journal*, 1(3), 3–9.

Zairi, M. (1996) *Benchmarking for Best Practice.* Butterworth-Heinemann, Oxford.

Zairi, M. (1997) Business Process Management: A Boundary-Less Approach to Modern Competitiveness. *Business Process Management*, 3(1), 64–80.

Zha, X. F, Lim, S. Y. E. & Fok, S. C. (1998) Integrated Intelligent Design and Assembly Planning: a Survey. *International Journal of Advanced Manufacturing Technology*, 14, 664–685.

Zirger, B. J. & Maidique, M. (1990) A Model of New Product Development: An Empirical Test. *Management Science*, 36, 867–883.

Index

Note: page numbers in *italics* refer to figures or tables